T0187147

Can Fish Count?

Can Fish Count?

What Animals Reveal About Our Uniquely Mathematical Mind

Brian Butterworth

QUERCUS

First published in Great Britain in 2022 by Quercus

QUERCUS

Quercus Editions Ltd
Carmelite House
50 Victoria Embankment
London EC4Y 0DZ

An Hachette UK company

Illustrations by Jeff Edwards

A CIP catalogue record for this book is available
from the British Library
HB ISBN 978 1 52941 125 6
TPB ISBN 978 1 52941 124 9
Ebook ISBN 978 1 52941 126 3

10 9 8 7 6 5 4 3 2 1

Typeset by Jouve (UK), Milton Keynes

Printed and bound in Great Britain by Clays Ltd, Elcograf S.p.A.

To Diana, Amy and Anna, who all count

CONTENTS

PREFACE

Modern science is a team sport. Who's in the team is partly a matter of luck, and I regard myself as being particularly lucky in the teams I've played with over the years. Not only couldn't I have written this book without chance meetings over the years with many people, I wouldn't even have thought about it.

One such meeting, at a conference in Ravello, was with Carlo Semenza, psychologist, psychiatrist and neuroscientist from Padua University in Italy. This led to a long collaboration, initially on disorders of language, but latterly on mathematical cognition and their disorders.

I probably wouldn't have started thinking seriously about numerical abilities without the initial prompt of my then-student Lisa Cipolotti. Lisa was one of Carlo's brilliant students, who asked to come to London to do a PhD on aphasia, but when she actually arrived she decided that rather than aphasia – which was very well studied – she wanted to work on a disorder very few people were researching. So we agreed to study the neuropsychology of mathematics, which at that time very few people anywhere were working on. Together with Carlo and his Austrian student

Margarete Hittmair, we formed a team with the pioneering neuro-psychologist Elizabeth Warrington at the National Hospital for Neurology in London, to investigate acquired disorders of mathematics, supported by a grant from the European Community. This also created a longer-term link between Padua and London that continues to this day.

Studying neurological patients revealed, first, that the key brain region for numerical processes was in a small part of the parietal lobe, and the adult brain network appeared to be independent of other cognitive processes (this was not a new discovery, but recapitulated in more detailed studies going back to the 1920s), but more interestingly it seemed to be organized into distinct components, each of which could, in some cases, be separately affected by brain damage. This was the adult brain, but I started to think that about how the brain develops these components and why these particular regions. Do we inherit a brain organized to extract numerical information in the environment? And if so, how deep into evolutionary history do these roots go? Can this inheritance go wrong, as in colour blindness?

In 1989, when Lisa first came to work with me, studies of numerical cognition were confined to separate silos with very high walls: neuropsychology of mathematical disorders, adult cognitive psychology, child development, animal studies, mathematical education, philosophy of mathematics and the beginnings of brain imaging. Practitioners in these silos rarely talked to each other, but I and a few others thought that the whole field would be advanced if they did. Then came another slice of luck. My friend Tim Shallice, at the International School of Advanced Studies in Trieste, found some money to fund a week-long workshop in Trieste in 1994, with which we were able for the first time to bring together many of the world's top scientists, and some of the world's top students, and give them the opportunity to talk to each other. As an almost immediate result, I managed to organize and find funding for a European network of six labs, Neuromath, and then a second network of eight labs, Numbra, to collaboratively pursue this interdisciplinary approach. Through the Trieste meeting and the networks, I have been able to meet, discuss and collaborate with an astounding number of inspiring scientists. Stanislas Dehaene was at the Trieste meeting and has been one of the most important shapers of this whole field, and his contribution has been fundamental to my own thinking. No high-class symposium would be complete without a philosopher, and we were lucky that my colleague at UCL, Marcus Giaquinto, was an outstanding philosopher of mathematics, and was able to keep us, me in particular, on the philosophically straight and narrow.

One student who was in Trieste at the time but wasn't officially at the Trieste meeting was Marco Zorzi. He spent some time in my lab doing ground-breaking work on using neural networks to model reading, and later to model basic arithmetic processes. Currently, Professor Zorzi in Padua runs one of the most innovative labs in the world for mathematical cognition.

Randy Gallistel and Rochel Gelman also were at the meeting, where we became friends, and I have since spent many happy hours in various parts of the world, often starting at breakfast, with Rochel and Randy, arguing about the nature of mathematical abilities in humans and other animals. Their approach to these issues, as you will see, has influenced me profoundly.

Randy, Giorgio Vallortigara, a brilliant animal experimenter at the University of Trento in Italy, and I organized a wonderful four-day meeting at the Royal Society in 2017 on 'The Origins of Numerical Abilities' that attracted an amazing group of scientists approaching the subject from many different perspectives, from archaeology to insects. Four days devoted to number – actually five, because the previous day Ophelia Deroy organized an international symposium on the philosophy of mathematics at the Institute of Philosophy in London. Five days on number: I was in heaven. In one sense, this book is an attempt to make the contents of these meetings available to the general reader.

Christian Agrillo from Padua, at the time a student, first got me interested in the numerical abilities of fish. I am currently collaborating with Caroline Brennan and Giorgio on a project about the genetics on numerical abilities of zebrafish. Tetsuro Matsuzawa, who I first met at a Neuromath network summer school, invited me to the Primate Research Institute of Kyoto University to observe his inspiring work with chimpanzees.

Fundamental to my whole approach has been my work with Bob Reeve on the early development of numerical abilities in mainstream Australia and with indigenous groups there.

My work has been supported over the years by many organizations and foundations. The Leverhulme Trust has supported our work with Aboriginal children, and our current fish study with Brennan and Vallortigara. The Australian Research Council has supported Reeve and me on our longitudinal studies of mathematical development. The Wellcome Trust has supported many of our studies with children, adults and neurological patients.

I must say a word of thanks to my literary agent, Peter Tallack of the Science Factory. He managed to get the project off the ground after years of my trying unsuccessfully to do it.

I had been interested in the foundations of mathematics since reading Bertrand Russell in my teens, and in particular Gödel's theorem, when I met, by luck, Diana Laurillard, a student of maths and philosophy, by gatecrashing her bonfire night party in 1967. Despite the brief interruption of a police raid, I learned that she was also interested in Gödel's theorem. The other piece of luck is that Diana remained interested in my work and in me until the present day. We now work together on how to turn evidence from the science into practical applications in education. She also patiently listens to, and corrects, my ideas. So some of the errors in this book have escaped even her close scrutiny.

CHAPTER 1

THE LANGUAGE OF
THE UNIVERSE

Galileo (1564–1642) said that the universe is written in mathematics and we cannot read it unless we become familiar with the characters in which it is written. Eight hundred years earlier, the great Persian mathematician Al-Khwarizmi (in Latin, *Algorithmi*) (*c*.780–*c*.850) wrote 'God set all things in numbers.' In 1960, physics Nobel Laureate Eugene Wigner (1902–1995) wrote a famous article entitled 'The Unreasonable Effectiveness of Mathematics in the Natural Sciences'. Mathematics, he said, had an uncanny ability to describe and predict phenomena in the physical world. The suggestion is that maths isn't just a tool to describe the world, but rather there is something about it that is profoundly mathematical. This is an idea that goes back even further than Al-Khwarizmi to Pythagoras (*c*.570–495 BCE), who allegedly said that all things are made of numbers.

In one sense, this is obvious nonsense, but perhaps there is a deeper truth here. Pythagoras may have been the first to observe the numerical structure of musical pitch, and we still use terms such as harmonic mean and harmonic progression. He also documented the relationship between numbers as shapes, and again we still use his terms: squares, cubes, triangular numbers and pyramidal numbers. Once we get into the mind of the Pythagoreans, it is possible to think of the world as being built, in a rather atomic or molecular way, out of these numerically defined objects.

Finding mathematical structures in the world is the work of people we would now call scientists, and it is them that Galileo, Al-Khwarizmi and Pythagoras were addressing. It has also been assumed that scientists in the rest of the universe, provided they were intelligent enough, would be able to read the language of the universe. If they wanted to show us that they were indeed intelligent, they would broadcast something numerical. The challenge of communicating with aliens by radio was taken up enthusiastically by Nikola Tesla (1856–1943), who claimed to have intercepted a signal from 'another world, unknown and remote'. It began with counting: 'One . . . two . . . three . . .'.[1] The American scientist Carl Sagan (1934–1996), in his science fiction novel *Contact*, had his extra-terrestrials sending a sequence of prime numbers.

In 1960, the Dutch mathematician Hans Freudenthal (1905–1990) published his Lincos code (*Lingua Cosmica*) that encoded not just numbers, by the number of pulses, but relations between them, such as equals to, greater than, and so on, to prove to the recipient intelligences that we were an equally advanced civilization.[1] In the movie version of *Contact* (1997), SETI (search for extra-terrestrial life) astronomers receive a radio transmission from space that has a Lincos-like dictionary embedded in the message.

But do you really have to be an advanced civilization, or even very smart, to have some understanding of the language of the universe? Can the relatively dumb fellow species of our own world also read the language of the universe, at least the type of characters proposed in Lincos, the whole or natural numbers 0, 1, 2, 3, and the relations among them?

In the physical world, whole numbers are fundamental: the water molecule has *three* atoms, *two* hydrogen and *one* oxygen; nitric oxide NO (an important cardiovascular signalling molecule) has *two* atoms, nitrogen and oxygen; nitrous oxide N_2O is an anaesthetic and has *three* atoms, *two* nitrogen and *one* oxygen; nitrogen dioxide NO_2 is a nasty pollutant, and has *one* nitrogen and *two* oxygen. We have *four* limbs, insects have *six*, spiders and octopi have *eight*. We have *two* eyes, but some spiders have *eight*.

These are real properties of the real world. Things would be very different if these numbers changed – for example, if we had three arms and three eyes. The mathematical structure of the world is important to us as scientists, but it could be important to other creatures too. Consider the following numbers in the real world.

I can see *three* ripe fruit on that tree and *five* on this tree. I can hear *five* invaders to my territory, but there are only *three* of us. There are *three* little fish like me over there, and *five* here. I can hear *five* croaks to a phrase over there and *six* near the lilypad. I passed *three* big trees between home and the food source.

All these numbers have an evolutionary significance, and if a creature can recognize them, this could afford an adaptive advantage. Foragers benefit from selecting the tree with five ripe fruit over the one with three, and the female frog benefits by mating with the male able to produce six croaks in a breath over the one that could only manage five (see Chapter 8). Lions are more likely to survive and reproduce if they only attack invaders when they outnumber them (see Chapter 5).

These ideas are the starting point for this book. Can our unique mathematical abilities have an evolutionary basis? How is it possible to tell if creatures without language can respond to the numerical structures of the universe?

In fact, there has been a hundred years of research into animal abilities to read the mathematical structure of their worlds. Now that doesn't mean that their abilities, if they exist, are evolutionary antecedents of our own. That would require a genetic link between us and them. There is an example that could serve as a clue. We know that there are genes for *timing* that have been conserved for more than 600 million years, from before the invertebrate line (insects, spiders and so on) separated from the vertebrate line (fish, reptiles and mammals). They are helpfully labeled CLOCK, PER (for period) and TIM (for regulating the time of the daily circadian rhythm). We can find these genes in fruit flies (*Drosophila melanogaster*) and the descendants of these genes from a common ancestor in humans. Timing is a mathematical property of the world because duration can be represented by a number. Thus we may find genes for numerical abilities to go with the timing genes.

What is number? What is counting?

Before delving any deeper I had better come clean about what I mean by numbers and by counting. All readers of this book think they know what a 'number' is. They may think of the words *one*, *two*, *three*, or of symbols – *1, 2, 3* – or both. Because we have been brought up in a numerate culture and have learned to count with counting words, we may habitually think of counting as necessarily reciting 'One, two, three . . .'. Scientists have to be more specific.

Of course, there are many kinds of number: positive whole numbers, sometimes called natural numbers; integers which include negative numbers; fractions, reals (decimals), imaginary (i, the square root of -1) and even the late, great John Conway's surreal numbers. Among the whole numbers are ordinals for ordered sequences, like pages in a book or house numbers in a street, which do not directly reference magnitude. So my house is number 44, but it is exactly the same size as my neighbours' in 42 and 46. This page is the same size as the next. There are also numerical labels for TV channels and phone numbers. These refer neither to the magnitude nor order. And it doesn't make sense to ask whether my phone number is larger or smaller than yours, nor to ask if it comes before or after yours. The type of number that does indicate magnitude is a *cardinal* number. These denote the size of a set.[2]

The idea of a set that underlies the idea of cardinal number needs some further explaining. Think of the set of three things – for example, the set of three coins in the fountain. Sets and their sizes do not depend on what the objects are: they can be three coins (physical objects), three knocks on the door (sounds) or three wishes (thoughts). The important thing is that these sets have nothing in common apart from their threeness. This leads to an ancient philosophical difficulty I will return to in the final chapter.

For the rest of the book when I talk about numbers I will mean cardinal numbers unless I specify otherwise. However, I want to introduce a new term, *numerosity*, to refer to set size rather than the logical and mathematical term *cardinality*. This is because we are talking about what goes on inside the brain of an animal, not about logic or mathematics.

I follow the eminent scholar Randy Gallistel in his proposal for assessing whether an animal, or a human, is actually capable of representing numerosity in their brains.[3] He sets out two criteria:

> (a) Do they represent numerosity as a distinct property of a set, separable from the properties of the items that compose the set?

This is exactly in line with what I have described.

It is not enough to represent numerosity; you also have to be able to do something with it. You have to be able to calculate, to do what Gallistel calls 'combinatorial operations' of a certain type.

> (b) Do they perform with the representatives of numerosity combinatorial operations isomorphic [equivalent] to the arithmetic operations that define the number system $(=, <, >, +, -, \times, \div)$?

Thus, we can ask whether the animal can in some way recognize that two sets have the same numerosity (=), and that set A is larger than set B (A > B), and that the sum of sets A and B is equal to set C (A + B = C). Division and multiplication you might think to be much more difficult, but for an animal to calculate how frequently food or a predator occurs this is a matter of division (e.g. 3 appearances per day = 24 hours/8-hour intervals). When it comes to navigation – 'dead reckoning' in sailing vocabulary – this involves quite complicated calculations.

Now these are fairly tough criteria. But I would elaborate (a) to ask *to what extent* can an animal *abstract* from a particular set to a novel set? In other words, can they assess whether sets with different types of object have, for example, the same or different sizes? For instance, can the animal notice that a set of sounds has the same numerosity as a set of food items? Other species, indeed some humans, may be able to apply numerosities only to some sorts of objects, perhaps those vital to one's life but not to other things. They may not be able to tell whether a set of one type of object, for example petals on a flower, has the same numerosity as other types of objects, such as landmarks, as I will try to explain a bit later.

Gallistel's two criteria reflect current philosophical thinking about the foundations of mathematics.[2] They also reflect how arithmetic is typically taught around the world: 1–1 enumeration of sets of objects, ensuring that enumeration does not depend on the nature of the objects (abstraction), and then working with the arithmetical consequences of operations on sets – comparing, combining and adding, splitting sets, subtraction, and so on.

WHAT IS COUNTING?

Most readers of this book, if they think about counting at all, will think of it as an activity that is intentional, purposeful, conscious and usually accompanied by the use of counting words. And an intention and purpose of this activity is establishing the *numerosity* of a set. This characterization rules out all non-human counting, and as we will see in the next chapter, some human counting. No other creatures have counting words – apart from the parrot Alex in Chapter 6 on birds – and ascribing intentions, purposes or consciousness to non-humans is, to say the least, controversial. We may be willing to ascribe them to the great apes or to pet dogs, but to fish or insects? No way.

Suppose you have to count the number of dolls on the table, and do this by counting out loud *one, two, three* dolls. This enables you to establish the size of the set of dolls on the table: three. Rochel Gelman and Randy Gallistel in their ground-breaking book, *The Child's Understanding of Number* (1978), listed the 'counting principles' that characterize human counting.[4] The 'cardinal principle' states that the counting process yields the cardinality of the set with the last word of the count, provided of course that all the objects are counted and each object is counted once and only once (1–1 enumeration). That is, there is a strict one-to-one correspondence between the counting words and the objects in the set. They also note that to have a sense of set size, it doesn't matter which object you start counting with; the size will always be the same. So for the set of three objects A, B, C, it doesn't matter whether you start with A, B or C, you will still end up with the numerosity of three. They call this the 'order-irrelevance principle'. Finally, they note that sets can be sets of anything, three dolls, three chimes, or three wishes. This they call the 'abstraction principle'. These principles are required for humans to be competent counters using counting words. They bring together the cultural tools – the counting words – with the concept of set size. I'll say more about how children learn to count with counting words in the next chapter.

Now consider one way in which we count the members of a set without necessarily using counting words, by using a very simple and cheap device called the tally-counter (see Figure 1). The button on the top is pressed once for each object counted – the one-to-one correspondence principle. The total of the count is given in the readout from the last button press – the cardinal principle. Anything that is countable can be counted (the abstraction principle) and the set can be counted starting with any member (the order-irrelevance principle).

To use the tally-counter a *person* also has to count. That's the hard part. Suppose you are a shepherd counting sheep but not goats – you have to be able to decide which is which. Then suppose you have to decide whether you have more sheep than goats – you will have to count the goats with a separate counter and then inspect the readout to see which number is larger. The tally-counter has a memory – a readout – that indicates the number of objects counted. Of course, you could use the same tally-counter twice, once for

Figure 1. Tally-counter

sheep and once for goats, but then you would need a memory for the sheep, set the counter to zero and start again for goats. In both cases, you also need a mechanism for carrying out the comparison. I'll show how this might all work in a moment.

The English philosopher John Locke (1632–1704) made an early attempt to characterize number and counting in a way that anticipates our tally-counter in *An essay concerning human understanding* (1690). He said the simplest idea is *one*; *one* can be repeated, 'and by adding repetitions together' – like repeating the button pushes – we get the complex idea of larger numbers. 'Thus, by adding one to one, we have the complex *idea* of a couple,' etc.[5]

This is an example of *recursion*, a procedure or function, that calls itself to do it again. Locke is proposing a particular form of recursion called 'tail recursion', where a procedure generates the last item (the tail) by adding one and then calls itself to carry out the procedure again, adding another one.

As to counting words, Locke says we should give each complex idea 'a name or sign, whereby to know it from those before and after'. The tally-counter can be thought of as an implementation of Locke's proposal. Each button press is a repetition of one, and the sum of button presses is given by a 'sign', in this case a readout in digital form.

The other issue is the degree of abstraction involved in numbers and counting. Big Ben chimes five times at five o'clock and we have five fingers on each hand, but chimes and fingers have no other properties in common. So how can brains, even tiny ones, deploy these abstract ideas, and how can we tell that they do so? We can ask humans to count out loud or say whether the number of fingers and chimes is the same, but we can't ask fish. To what extent can other creatures generalize the numerosity in one modality – sounds, for example – to numerosities in other modalities – visible objects, actions and so on? This will be a matter for a *selector* rather than the counter. Although I have given this component a special name, it is really based on widespread ideas in cognitive science and neuroscience. It is a way of focusing on or attending to an object or an event. The selecting process does not have to be conscious or even intentional, which are controversial concepts when applied to non-human animals. It simply picks out one or more objects from the environment for further action.[6] We can use a single tally-counter to count the chimes and the fingers, as long as it can access two memory locations.

A NEURAL COUNTING MECHANISM

Of course, we don't literally have tally-counters in our heads, but do we have some kind of neural equivalent? The tally-counter is a *linear accumulator* with a memory. That is, the contents of the accumulator is strictly proportional to the number of objects.

That our brains possess such a mechanism is an old idea that actually comes from work with animals. The same accumulator mechanism can also measure duration which is needed when animals need to calculate the rate or frequency of events.[7]

The accumulator mechanism needs four components:

- *An internal generator*, like an oscillator or *pacemaker*, of which there are many in the brain, which sends pulses at regular intervals to an accumulator.
- *A normalization process* that treats all objects or events as equivalent.

- *A gate* to control the transmission of the pulses between generator and the accumulator. When a selected object or event is to be counted, the gate is opened to allow a fixed number of pulses into the accumulator.
- *An accumulator* that temporarily stores the pulses.

As well as these components, the accumulator counting system needs a *working memory* to store the results of the current count, and a *reference memory* for future use. To go back to our shepherd, the number of sheep is transferred to the reference memory, and the current count of goats to the working memory. Figure 2 is one way of representing the accumulator.

There is, as I noted, a further component that is neurally much more expensive: the selector. This chooses the objects to be counted, like sorting sheep from goats. There also needs to be a

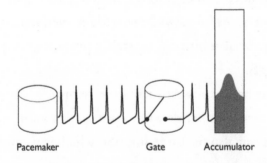

Pacemaker   Gate   Accumulator

Figure 2. This is the earliest version of this model by the animal psychologists Warren Meck and Russell Church.[7] They used the term 'pacemaker' and the 'gate' for normalized objects or events by allowing a fixed number of pulses for each object or event to go through to the accumulator. Here I show each object as two pulses. To carry out combinatorial operations that are equivalent to arithmetic there will need to be a 'working memory' to store the state of the accumulator temporarily plus a 'reference memory' so that the operations, including comparison, can be carried out (<, >, =, +, −, ÷, x).

mechanism for carrying out 'combinatorial operations' – for example, a combinatorial operator for deciding whether the set of sheep is larger than the set of goats (isomorphic with <, > or =), or the total number of sheep plus goats (addition), or exactly how many more or fewer there are of sheep than goats (subtraction).

Meck and Church proposed that the same accumulator mechanism could also be used in a *continuous mode* to measure duration. The gate is held open for the duration of the event so that the content of the accumulator is linearly proportional to the duration. Thus both numerosities and events are coded in the same way as contents of the accumulator, and continuous quantities such as duration are coded in the same way as a level in the accumulator. Because numerosity and duration are coded in a 'common currency', the accumulator level, it is possible to compute parameters important to the organism such as the rate or the probability of an event occurring (duration/number).

This is a very simple device, like the tally-counter, that needs few neurons to implement, as we will see when I come to the counting abilities of insects and spiders, which have very tiny brains (fewer than 1 million neurons compared with our 85+ *billion*). The expensive element in the system is the selector, as in the sheep and goats example. The selector also has to decide whether it is something other than a sheep or goat and whether it is actually a single object (e.g. not a lamb suckling its mother's teat).

The linear accumulator has several attractive features. It meets Gelman and Gallistel's three counting principles: abstraction – anything the selector can identify can be counted; order-irrelevance – it doesn't matter which object is counted first, second and so on; and cardinality – the final level of the accumulator is the total of the count.

Accumulators also meet Gallistel's second criterion of the representation of number: that operations on accumulators or on memories of their output are isomorphic with arithmetical operations: comparing levels of two or more accumulators or the reference and working memories (=, <, >); combining is the linear sum of the accumulators (+); and subtracting is the linear result of taking one accumulator from another (–). Division and multiplication come into play when computing rates or probabilities, which we will see that animals can do.

An alternative view

Although many scientists subscribe to a version of the accumulator theory, currently the most popular approach in the scientific literature on basic numerical abilities in humans and other animals postulates two separate systems, one for small numerosities (≤4) and one for larger numerosities (>4).[10] In the two-system account, the first system is premissed on a perceptual system designed to track objects and briefly remember them, called the 'object tracking system' (OTS). The limit of four items corresponds to the number of objects one can attend to at any one time and the corresponding working memory of them, and the limit of the numerosity of objects that can be accurately and speedily recognized without counting. It is assumed that this mechanism operates with 'parallel individuation' of the objects in this small set; that is, it does not serially scan the array of objects and then add them to a temporary store, but rather takes them all in at once. This is sometimes called subitizing (from the Latin *subitus*, meaning 'sudden'), and implies that it is as easy and as fast to recognize the numerosities of all sets from one to four.[11]

The second system for numerosities greater than four is called the 'analogue magnitude system' or the 'approximate number system' (ANS). The characteristics of this system are, first, that the mental representation of a numerosity is approximate, so that, for example, the representation of five objects will to a lesser extent also include the representations of fourness and sixness, and to an even lesser extent, threeness and sevenness. Second, the representations are logarithms of the numerosities.

This alternative approach has been challenged in several ways. The assumption that there are two separate systems has frequently been disputed. For example, the response time to name the number of object in the subitizing range (≤4) actually depends on the number. It's not just 'sudden' recognition of numerosities one to four. Rather, there is an increase in the time to recognize the numerosities of randomly arranged dots briefly exposed: an increase of 30 msecs from one to two, 80 msecs from two to three, 200 msecs from three to four, and thereafter an increase of 300+ msecs per item.[11] One implication is that the reaction time can be fitted with a single curve consistent with a single mechanism.[12] In fact, there is evidence from monkeys (Chapter 4) and humans of a continuity of numerosities from the small number range right up to thirty.[13]

The other argument against a separate subitizing mechanism is that no one has found distinctive activations in the brain, and we tried really hard to find them in the parietal lobes.[14] That was twenty years ago, but a recent study with much more precise high-resolution scanning and better analysis tools found that activations for all numerosities from one to nine were located interspersed in the same brain regions, not just the parietal lobe but also visual (occipital) and frontal cortices.[15] This doesn't mean that there aren't neural differences, only that they have yet to be identified, which may be very difficult given that our most powerful imaging tools may still be too crude to pick out the differences.

The other important claim for the ANS is that the internal representations are *logarithms* of the numerosities. Elizabeth Brannon and Dustin Merritt point out that these two models 'predict the same behavioral signatures' on tasks involving ordering numerosities, for example in picking the larger or the smaller because the 'noisiness' (variability) of the representations is similar. 'To adequately differentiate the logarithmic and linear with scalar variance hypotheses, it is necessary to use a task that required subjects to base their behavior on the *difference* between two points on a continuum,' they argue.

Brannon and her colleagues used this principle in a study with pigeons. Their finding is consistent with a linear internal representation.[17] Slava Karolis, Teresa Iuculano and I, in a study of humans, have used the differences between numbers on a scale to 100 and also found evidence in favour of a linear internal scale.[18]

Of course, the other problem with logs of numbers is that it is hard to do simple linear addition and subtraction without an antilog table since log A + log B = log AB. As we will see in later chapters, there is clear evidence that many animals can add and subtract. Do animals – including us – have an antilog table in our brains?

Now it is true that both adults and especially children look as though they have a logarithmic mental number line because large numbers are underestimated and therefore compressed together and small numbers are overestimated with the result that they are more spread out.[19] However, this compression does not mean that the mental scale is logarithmic. Indeed, it has been known for more than a hundred years that our judgements have a 'central tendency' to overestimate small values and underestimate large ones for all kinds of things without having a logarithmic mental representation of the scale.[20]

What about really large numerosities when there isn't enough time to count all the objects? A ground-breaking study of sub-itizing by George Mandler and Billie Jo Shebo implied that beyond about ten objects a completely different system is used.[11] Subsequent researchers have suggested that for these large visual arrays, we use a method of *estimation* based on area and density. That is, if the objects, such as dots, cover a large area or are densely arrayed, then the observer – human or other animal – does something like multiply area by density to get the estimate of numerosity. This seems to me entirely plausible. The estimate may then be mapped on to an internal representation, such as accumulator height or location on an internal number line, to get 'about 30' or 'about 100'.

Methodologies for study of other species

There are two main methods for testing whether a creature, human or not, can represent the number of objects in a set. The first is 'doing what comes naturally'. This can be explored in the lab, and it can be observed in the wild. For example, one can offer the creature two bits of food and three bits of food and see if it selects the larger. This method is sometimes called 'going for more', which is what we assume the creature will do both in the wild and in controlled conditions in the lab. The second tests whether the creature can learn to select on the basis of numerosity even if it doesn't come naturally – for example, learning to pick the display with more dots to get a reward.

I'll now briefly outline some of the basic principles for applying these methods.

The Clever Hans problem. Let me start with the famous example of an animal which, despite appearances, did not count: the horse Clever Hans. He flourished at the turn of the century and was able to produce answers to arithmetical problems by tapping his foreleg. Some of these problems would be too difficult for many high school graduates. He was able to add fractions such as ⅔ and ½, and have the answer by tapping out the numerator, 9, and the denominator, 10, separately. He could find factors, for example, of 28, and tapped out correctly 1, 2, 4, 7, 14 and 28. He could give square roots and cube roots. Extraordinary. Obviously, people thought fraud must be involved; but it wasn't. He was tested by a panel of psychologists and animal trainers, but none could find evidence of fraud. To cut a long and very interesting story short, Hans was indeed clever, but not at arithmetic. He was very sensitive to cues given off by the tester, not only by his trainer but even by members of the panel brought in to investigate. Hans was cued to stop tapping by the movement of the tester's head, eyebrows, or the dilation of his nostrils as Hans approached the correct number. Oskar Pfungst, who finally discovered the cueing, found himself unable to stop providing the cues when he tested Hans. Subsequent research has tried to avoid this problem by keeping the experimenter concealed from the animal. (For a nice account of this issue, see a report from animal behaviour expert Hank Davis[21]).

The first question in many animal experiments is: does the animal respond to (or notice) difference in numerosity, taking everything else into account? In nature, changing the number of objects changes lots of other non-numerical features. 'Everything else' will include the total amount of stuff, such as the amount of blackness if it is black dots, or the amount of fishness if it's fish; but it also includes sizes of black dots and how densely they are arranged.

Here's the problem: if you want to know whether the observer notices the change from one black dot to two black dots (and the dots are the same size), then there will be twice as much blackness for two black dots. If you then try to control for the amount of blackness by making the two black dots half the area of the single dot, the observer, human or not, may be noticing the difference in the size of the dots, or the total length of the edges of the dots. Unfortunately, too many researchers think that controlling for the amount of stuff does the job. It doesn't. There are various ways round this. One way is to randomly change size, area and density from trial to trial. This works if the sequence is quite long. Another approach is to change the objects completely, for example from dots to squares, but here you have to take into account that the viewer may be noticing both the change of object and the change of number, and this necessitates quite a long experiment to sort it out statistically.

The third way is to use a method called 'match-to-sample'. In this you present the subject with a sample of, say, two dots and the task is to find another set of two dots, again trying to control for co-occurring changes.

The ground rules for this method with animals were laid down by the German ethologist Otto Koehler (1889–1974), who worked mostly with birds, but also with other species. He made the sample very different from the choices: see Figure 3 (overleaf).

The match-to-sample method has been used with many different species, though rarely with humans. Koehler also used another way of doing match-to-sample. He would allow the bird

Figure 3. Match-to-sample in corvids. The bird is shown a sample (the larger square) and is rewarded by locating the lid with same number of objects, here dots of various sizes.[22]

(or squirrel: he tested these also) to see or touch a sequence of boxes, some with a bait in them, and then match the number of baits collected to a box with same number of dots. That is, the animal counts n baits and is then rewarded by selecting another box with n dots on its lid. This is a fuller abstraction than matching of dots on two lids as in Figure 3.

One can go even further with this method, for example by

matching the number of sounds to the number of objects, or the number of actions. In our study of Aboriginal children in Australia, the task was to match the number of sounds made by hitting two sticks together by putting out the same number of counters on a mat (see Chapter 2). This kind of cross-modal matching is a powerful test of representing the numerosity of sets, and doing so in a very abstract way.

I will describe variations on these two basic paradigms in subsequent chapters.

Weber's Law and the Weber Fraction. This is a very important law that crops up innumerable times in the discussion of numerical processing. Weber's Law is mainly relevant to whether an animal can *notice* the difference between two numerosities. The German physician Ernst Heinrich Weber (1795–1878) was perhaps the first to discover that our ability to accurately discriminate between two quantities – he started his research with the discrimination of weights – did not depend on the absolute difference between the two weights but on the proportional or ratio difference between them. That is, it is easier to pick the heavier of 1 vs 2 kilos than 10 vs 11 kilos. It is also easier to pick out candlelight in a dark room than in bright sunlight. And I wonder if years seem to go faster when you're older because each additional year is a smaller ratio of all your years.

These observations led Weber to formulate his now eponymous law: $\Delta I/I$ is a constant, where I is the reference value of, in this case, weights, and ΔI is the difference in weight. So, for numbers, if you have to choose the larger of 3 and 4, the absolute difference is 1, but the incremental ratio difference is ¼, or 0.25. If you have to choose the larger of 13 and 14, the absolute difference is ¹⁄₁₄ or 0.07, much smaller, and therefore a more difficult decision. The value of the constant – the *just noticeable difference* – will depend on many factors, including individual differences. Some individuals will have better discrimination than others, and this will turn out to be very important (see Chapter 2 on humans and Chapter 8 on fish). For example, if my personal Weber fraction for numerosity comparisons is about 0.20, I'll have no trouble distinguishing 1 vs 4 (0.75) dots or 3 vs 4 (0.25) dots, but I will struggle with 13 vs 14 dots (0.07).

Now when making a discrimination, there will be other factors to take into account. The brain is noisy. This means that with two weights or two numerosities, A and B where objectively A>B, you will sometimes judge that B>A, especially if the ratio difference between A and B is close to your just noticeable difference.

In fact, noisiness increases with the magnitude – the higher the number, the more the curves are spread. This feature of brain activity is called 'scalar variability' – that is, the larger the number, the more likely the error and the larger the error is likely to be. Errors scale precisely in proportion to the magnitude of the number. To put it more technically, the coefficient of variation (standard deviation/N, where N is the number) is a constant.

Our world of numbers

I once decided just for fun to see just how pervasive numbers are in the modern world, using myself as a subject. In the course of my work as a scientist, and particularly as a scientist working on how the brain deals with numbers, naturally I would be experiencing far more numbers than the typical citizen. So I assessed how many numbers I would experience on a Saturday when I wasn't working but just reading the paper, going for a walk, listening to the radio and doing a little light shopping. There were car numbers, parking signs, a postcode (zip code) on every street corner, bus numbers on buses and bus stops, and prices in shop windows. Of course, the newspaper provided a lot – page numbers, dates, financial information and many numbers on the sports pages, which I read avidly. Listening to the news on the radio meant even more numbers about weather, deaths, business and sport. I calculated that I experienced about 1000 numbers every waking hour.

I thought I would check this number again on a Saturday with a similar shopping route, about 1 mile there and 1 mile back.

I also get *The Economist* on Fridays and read it mostly on Saturdays. Lots of numbers in that as well. During the Covid-19 pandemic, we were daily assaulted by even more numbers: infections, deaths, vaccinations, costs. In the happy, or unhappy, event of an election, the numerical assault is almost overwhelming. Think of the CNN coverage of the US presidential, House and Senate elections in 2020. Numbers by state, by district, by exit polls for each of these contests; numbers as the counts come in; projections of the final score. Before the election there were more numbers: daily polling results leading to forecasts of the final outcome.

There was no escape from numbers on my brief shopping trip. I probably saw about 200 parked cars in two miles – OK, I didn't actually count them, but all kerb parking spaces were fully occupied. At two large digits per car, that's 400 digits, plus about 50 moving vehicles, that's another 100. At each corner there is a number designating the postal district. Say another 25 digits. On the way there and back there were road signs indicating parking times, speed limits and numerical restrictions of various types. Then there was shopping; just three items were bought, but outside every shop there was a sign indicating its phone number plus other information, and special offers. My walk took me past a cemetery and signs memorializing the celebrated dead. Even in death, there is no escape from numbers.

Overall, I probably experienced far more than 1000 numbers per hour reading the paper and shopping, but the rest of the day was more relaxed with a friend for lunch, but more numbers discussed. The costs of redecoration, the time of this and the time of that, as well as very visible digital clocks and thermometers for the inside and outside temperatures. Radio and TV provided many more – channel idents, current times, programme times, episode numbers and of course the news.

My guess was still about 1000 numbers per hour when not working. That's about 16,000 per waking day, or about 6 million per year, not counting dreaming numbers, or numbers when working, whether as a scientist, a shop assistant at the cash register, a shelf-stacker or a banker.

Most of the numbers I experienced when shopping were entirely irrelevant to me, apart from deciding on the purchases. Most of the numbers in the newspapers were of marginal interest as well. Nevertheless, unattended and even unconscious numbers are registered by the brain even when they are irrelevant or even if we are actually not aware of them.[23] I describe an experiment we did that demonstrates this in Chapter 2.

Numeracy is important for a numerate society

It's not just the number of numbers that's important, it is how we understand them. Poor understanding of numbers is a serious handicap for individuals and a major cost for nations. It makes individuals less employable, creates a risk of depression in adult-hood, and lowers lifetime earnings significantly.[24] In the UK about 25 per cent of adults have poor functional numeracy – that is, 15 million adults are estimated to have numeracy skills lower than those expected of an 11-year-old. Of these, 6.8 million have skills below the standard expected for a 9-year-old.[24] These problems persist into adulthood: 74 per cent of 37-year-olds have problems with division, 57 per cent with subtraction, 15 per cent could not manage their household accounts and 8 per cent could only manage their household accounts with difficulty.[25] Low numeracy causes poorer educational outcomes, lower earnings and trouble with the law, and impacts on mental and physical health.[26] It is a cause of distress, low self-esteem, stigmatization and disruptive behaviour in class.[27] A recent report stated that the cost of adult low numeracy 'can be counted in lost earnings – the £25 billion . . . would be added to our collective pay packets if numeracy skills could be levelled up'. That is about £1700 per person per annum.[28]

Indeed, low numeracy can be a matter of life and death. A large-scale study of adults with colorectal cancer in the UK and US found that those with low numeracy were less likely to intend to participate in screening and were more likely to be defensive in getting cancer information, and hence more likely to be untreated or treated too late.[29]

Low numeracy is also a cost to nations. In 2009, it was calculated that the cost of the lowest 6 per cent of numeracy to England is about £2.4 *billion* a year in lost taxes due to lower earnings, higher rates of unemployment, and increased costs of crime, social security, education and health.[25] At today's prices, the number would be higher.

A more serious numeracy problem is developmental dyscalculia, which affects 4–7 per cent of children.[30] Child and adult sufferers have extremely poor numeracy and struggle with simple numerical tasks such as remembering PINs or times tables, telling time and calculating journeys. This condition persists into adulthood and needs specialist help. Dyscalculia has even more profound implications for the individual than the low numeracy described above. A major government report in 2008 stated that:

> Developmental dyscalculia . . . can reduce lifetime earnings by £114,000 and reduce the probability of achieving five or more [acceptable grades in a public exam] by 7–20 percentage points.[31]

Needless to say, the incoming Conservative government ignored this report and the plight of dyscalculic sufferers for 13 years. In 2020, the government's chief scientific adviser wrote to the prime minister, Boris Johnson, to recommend recognition and intervention for dyscalculia. I am not holding my breath.

Worries about the state of the nation's mathematical competence in the UK – especially basic arithmetical competence – go back at least to the nineteenth century. In his foreword to the Cockcroft Report on mathematics education in 1982, Sir Keith Joseph, then secretary of state for education and science, wrote that 'Few subjects are as important to the future of the nation as mathematics.'[32] Since Cockcroft, there have been two more major reports. Similarly, the US National Research Council noted that 'The new demands of international competition in the 21st century require a workforce that is competent in and comfortable with mathematics.'[33]

The link between numeracy and economic development was clearly established in an OECD modelling exercise. It showed that the level of basic numeracy is a causative factor in a nation's long-term economic growth.[34]

So how well we humans who live in numerate societies deal with numbers is important both to ourselves as individuals and to our communities. In the next chapter, we'll see that humans living in non-numerate societies also have a mechanism for reading the language of the universe that can be revealed by appropriate testing.

Looking ahead

We now have criteria – representing the numerosity of sets, and carrying out arithmetic operations on these representations – for telling whether an animal, human or not, can represent and calculate with numerosities, and there is an outline of the two main methods of finding out if the animal meets those criteria, and with what limitations.

I will also try to explain *why* numbers are important to animals. We used to be taught that animals had various 'drives' that they try to satisfy. Which drive is dominant will depend on current circumstances. Most animal studies are about the drive to reduce hunger, since this is easiest to control in the lab. We'll see in Chapter 9 that cuttlefish behaviour depends on whether it's hungry or not. If it's hungry it will go for one big prawn; if it is satiated it will go for two small prawns. However, we can observe the role of number in the drive to avoid danger or death, and the drive to mate.

Although we now know much more about numerical abilities

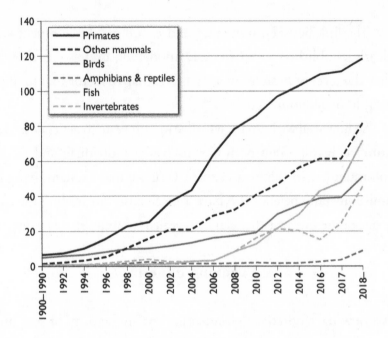

Figure 4. The graph shows the number of results returned from a search for ('numerical abilities' + animals) and a search for ('numerical competence' + animals) on Google Scholar.

in animals including human infants and adults than we used to, there are still vast gaps in our knowledge. Some animal classes have been more investigated than others. Christian Agrillo and Angelo Bisazza from Padua University, Italy, have a useful summary of the most and least investigated classes up to 2017 as far as numerical abilities are concerned. I have updated their figure to 2021 and included invertebrates (Figure 4).

There are many reasons for the discrepancy among species. One reason is that a particularly brilliant scientist has shown how rewarding studying a species or group of animals can be. Another is that the more is known about a species, the more new research can build on those findings. The primacy of primates is due to the fact that they are more closely related to us than other species, and it is believed they can tell us more about our own abilities.

There is another serious limitation of our knowledge of the numerical abilities of animals. Rarely do studies explore the upper limits of accurate counting; rarely do they explore the kinds of calculation that the organism finds possible; and, as will become clear, not all take into account the other cues that the animal might have used in the task. Not all experimenters follow Koehler as strictly as needed to draw conclusions.

Finally, there are extraordinary feats of computation that I will argue have to involve numbers but are still not well understood, and that is in animal navigation. We know that birds, whales, turtles, fish and even invertebrates undertake extraordinary journeys between their foraging and breeding grounds. These fundamentally depend on these creatures having at least a map and a compass, but it also means they have to measure distances to locate where they are and how to return by the shortest route. Think of how Google Maps encodes map information as an array of numbers, ultimately 0s and 1s, and computes routes over those numbers. These animal navigators have to have something equivalent to Google Maps both to represent their environment and to compute routes in it.

Being able to read the numerical language of the universe is vital for non-human animals, as life, death and reproduction all depend on this ability. And it is important for us to understand that our own extraordinary numerical abilities are founded on a simple mechanism that we share with many other, perhaps all other, creatures.

CHAPTER 2

CAN HUMANS COUNT?

The language of the universe is mathematical and being able to read this language is useful and indeed adaptive for both human and non-human inhabitants. I suggested that we and other creatures possess a very simple mechanism, the accumulator, that enables enumerating sets of objects or events.

So an obvious first question is, given these proposals, can humans count? Maybe not all of them, according to Dave Barry in the *Washington Post* (25 August 1996). 'They discovered that the bees were locating the feeder by COUNTING THE LANDMARKS. Yes! Bees can count! This means that bees, in terms of math skills, are ahead of most American high school graduates.' OK, the *Washington Post* is not a peer-reviewed scientific journal, and no data from high school graduates are presented, and this is unfair to American students. You and I can count. Everyone you know can count, even American students. In fact, there are some amazing human counters with extraordinary numerical skills.

In Chapter 1, I argued that counting is only relevant and only makes sense when the results of a count can be used in a combinatorial process that is isomorphic with arithmetical operations. This is true for us humans, and as I will show, for other animals as well. So when we ask whether humans can count, we are also asking what humans can do with the results of counting, namely arithmetic.

Most of us have heard of individuals with remarkable numerical skills – for example, Kim Peek (1951–2009), the model for Dustin Hoffman's character in the movie *Rain Man*. Others will have encountered the extraordinary recent feats of Rüdiger Gamm (born 1971), who taught himself to calculate numbers with very high powers in his twenties to win a prize on a German TV show. He takes less than five seconds to solve tasks such as 68 x 76 = ?. For me this would take seven steps with six intermediate results that I would have to hold in memory or write down. (Two-digit squares, such as 68^2, take him just over a second because they are simply retrieved from memory.)[1]

Nowadays, there are even mental calculation world cups. These include finding the square roots of six-digit numbers, calendar calculations ('What day was 3 January 1649?'), multiplying two eight-digit numbers, and so on. (Gamm only came fifth when he participated!)

In fact, there is a long history of individuals with remarkable numerical skills. Invariably they develop a kind of intimacy with numbers from an early age. When George Bidder (1806–1878), an exceptional calculator and a leading engineer of his time (a collaborator with the locomotive engineer Robert Stephenson) was learning to count to 100, the numbers became 'as it were, my friends, and I knew all their friends and acquaintances'.[2] Another exceptional calculator, Wim Klein (1912–1986), said 'Numbers are friends for me . . . It doesn't mean the same for you, does it,

3,844? For you it's just a three and an eight and a four and a four. But I say, "Hi, 62 squared." '

All extraordinary calculators have in memory an enormous store of number facts. Take the New Zealand mathematician Alexander Aitken (1895–1967): for him the year 1961 evoked the thoughts 37 x 53, $44^2 + 5^2$, and $40^2 + 19^2$. He could also recite the first 100 decimal places of π.[3] Why did he learn all this stuff? For Aitken, a teacher 'chanced to say that you can use factorization to square a number: $a^2 + b^2 = (a + b)(a - b) + b^2$. Suppose you had 47 – that was his example – he said you could take b as 3. So $(a + b)$ is 50 and $(a - b)$ is 44, which you can multiply together to give 2200. Then the square of b is 9, and so, boys, he said, 47^2 is 2209. Well, from that moment, that was the light, and I never went back.'[3]

In a famous story, the eminent English mathematician G.H. Hardy visited Srinivasa Ramanujan (1887–1920), whom Hardy regarded as the greatest mathematician since Gauss. He mentioned that the taxi in which he had come was number 1729, 'a rather dull number'. 'No, Hardy! It is a very interesting number. It is the smallest number expressible as the sum of two cubes in two different ways.'[4]

Perhaps even more extraordinary are the number skills of many children in Japan, China, Taiwan and India, who are subjected to extensive abacus training, usually in after-school classes. This can involve many years and hundreds of hours of deliberate practice, often directed towards success in competitions. After a while, an actual physical abacus is no longer needed, indeed is a handicap. Experts use a mental abacus. One type of competition is called 'flash anzan', where competitors must add numbers presented at a rate where it is scarcely possible to read them, let alone remember and manipulate them. Here's an example, taken from Alex Bellos's book *Alex in Numberland*. Children looked at a screen. After three beeps to alert them, the following numbers appeared so fast that Alex, an expert mathematician, could barely read them:

164

597

320

872

913

450

568

370

619

482

749

123

310

809

561

As soon as the last number flashed by, one student announced the answer, 7907.

The 2012 world champion, Naofumi Ogasawara was able to add correctly fifteen four-digit numbers presented at 0.4 seconds each.[5] As far as I know there has been no scientific study of how these amazing feats are achieved, in terms of either cognitive theory or brain function.

One fascinating feature of the ability to use a mental abacus is that you can carry out another task in parallel, provided it does not involve numbers. In a clip from Alex's YouTube channel, nine-year-old girls are doing a challenging language game while adding a sequence of thirty rapidly presented three-digit numbers in twenty seconds.[6] This suggests that calculation is a separable mental process from other aspects of cognition.

Now, you may think that these extraordinary numerical feats depend on great natural ability, for example, of intelligence or memory. After all, Ramanujan, Aitken and Bidder were indeed exceptionally talented.

Not all great calculators are talented or even especially intelligent. Consider Shakuntala Devi (1929–2013), for example, who was in the *Guinness Book of World Records* for being able to multiply two thirteen-digit numbers in twenty-eight seconds. She was rigorously tested by Arthur Jensen, a psychologist expert in intelligence testing, and was found to have average intelligence. Alfred Binet (1857–1911), the originator of the first practical intelligence test, compared the performance of two professional theatrical calculators – that is, people who made their living demonstrating calculation skills in theatrical performances – with cashiers from the Bon Marché department store in Paris, who each had fourteen years' experience of calculating (there were no mechanical calculators available in the 1890s), but who, presumably, showed no special early gift for mathematics. The cashiers did better.[7]

In fact, there are many examples of people with extraordinary numerical abilities who seemed to have had ordinary or even very low cognitive general ability. Zacharias Dase (1824–1861), who did table calculations for the great mathematician Carl Friedrich Gauss (1777–1855), was 'unable to comprehend the first elements of mathematics'. One pair of twins with prodigious abilities for calendrical calculation were estimated to have IQs in the sixties (the average is 100) and had great difficulties with simple arithmetic.[8] In this book on great calculators, Steven Smith noted an early report that two prodigious calculators, the enslaved African Thomas Fuller (1710?–1790) and Jedediah Buxton (1702–1772), 'were men of such limited intelligence that they could comprehend scarcely anything, either theoretical or practical, more complex than counting'.[2]

Henri Mondeux (1826–1861), a famous calculator, was described by a contemporary as never having learned anything besides arithmetic; 'Facts, dates, places, pass before his brain as before a mirror without leaving a trace.'[2]

Zerah Colburn (1804–1839) was able at the age of six to calculate the number of seconds in 2000 years (63,072,000,000) but was 'unable to read and ignorant of the name or properties of one figure traced on paper'.[9] Even as an adult, 'he was unable to learn much of anything, and incapable of the exercise of even ordinary intelligence or of any practical application'. In his 1891 review of calculating prodigies, the American psychologist Edward Scripture inferred that 'calculating powers . . . seemed to have absorbed all his mental energy'.[9]

These examples show that with deliberate practice, perhaps even more than the recommended 'ten thousand hours' to achieve expertise, humans can become remarkably skilled at numerical tasks. They also suggest that the numerical mechanism in the brain is independent of other cognitive mechanisms. Even the memory systems for number facts seem to be specialized for them. Gamm, for example, was able to repeat accurately eleven spoken digits whereas controls, and most of us, can manage to repeat accurately only about seven. If asked to repeat the digits in reverse order, he was even more remarkable: twelve digits backwards compared with five to six for the rest of us. However, ask him to do the same with letters, and he was entirely normal. And as we have seen, the memory for non-numerical information, in some of the extraordinary calculators, was very poor indeed. All this points to at least a partially separate cognitive system for numbers.

In fact, these skills, once acquired, can be deployed unconsciously, and counters are not aware that they have counted. But we, the scientists, know that they've counted, because it has a measurable effect on a task they are aware of. The way my colleagues Bahador Bahrami, Geraint Rees and I, and three brilliant students from the European Union on the Erasmus programme, showed this was the case by taking advantage of a phenomenon known as 'interocular suppression'.[10] The suppressing stimulus was a brightly coloured abstract pattern presented to one eye while the objects to be counted were presented to the other, suppressed, eye, and then both eyes received the abstract pattern. It is critical to get exactly right the interval between the objects to be counted and the suppressing stimulus: too long and the participant will be aware of the objects; too short, and the information will not register on the brain.[10]

Obviously, if they can't report the number of objects, how do we know whether they have counted them? We presented participants with another similar set of visible objects and ask them to count them. We found that if the suppressed set is smaller than the visible set, the visible set is counted faster and if it's larger than the visible set, it is counted slower. That is, the numerosity of the suppressed set affects the subsequent enumeration of the visible set by priming or inhibiting the decision. What this shows is that the counting process does not need to be conscious.

Many of us believe that we count in our sleep. Now ingenious experiments by an international team show that not only do we count in our sleep, we calculate as well.[11] In REM (rapid eye-movement) sleep, sleepers tend to dream and to be more sensitive to external stimuli. In this study, subjects were trained to link a sound to a task, and to respond by moving their eyes alternately left and right a certain number of times, for example, LR = 1, LRLR = 2, LRLRLR = 3, LRLRLRLR = 4. Simple addition and subtraction problems were presented during REM sleep. Amazingly, many of the sleepers answered correctly for at least some of the trials.

Learning to count with words and symbols

We have all learned the counting words, *one*, *two*, *three*, and *1*, *2*, *3*, but learning to count with these words and these symbols is actually not a trivial matter. Karen Fuson showed not only that it takes several years to completely master them, but there are distinct stages for in learning the counting words, beginning with thinking that counting sequence 'one two three four' is a single word, but eventually being able to go through an extended sequence forward and backward.[12]

To start with, we humans have to learn number words. These are cultural tools that have developed over centuries, perhaps millennia (see Chapter 3), and some tools are harder to learn than others. For example, in English there is the 'trouble with teens'. I mean, how could you understand *eleven*, *twelve*, *thirteen*, *twenty*, *thirty* on the basis of knowing *one*, *two*, *ten*, and *three*? Other European languages are no better. Take French: *onze*, *douze*, *treize*, *quatorze*, *quinze*, *seize*; or Swedish, where ten is *tio*, but eleven is *elva*, twelve is *tolv*, thirteen is *tretton*.

Contrast this with Chinese: one is *yī*, two is *èr*, three is *sān*, ten is *shí*, and eleven is *shíyī*, twelve is *shièr*, twenty is *èrshí*, twenty-one is *èrshíyī* and you are now in a position to fill in thirteen, thirty-one, thirty-two and thirty-three.

The Chinese system, which is also the basis of Korean and Japanese counting, makes the decade structure clearer.[14] Now, the counting words are a name-value system in all these languages; that is, each decade or century has a special name – *ten, hundred, thousand, shí, bǎi, qiān* – but it is easier to learn the link between the names and the digits in Chinese, Japanese or Korean than in European languages, and first-year schoolchildren think in terms of tens and units much earlier than English-speaking children in the US.[15] East Asian children also understand the base-ten system better than US, French and Swedish first-graders.[16]

To count with counting words, you need to know the words are in a stable order, and eventually the correct order. As Fuson discovered, some kids start by thinking that *onetwothreefour* is a single word; other kids know some of the words in the correct order, but not all of them. Here's an example from Rochel Gelman and Randy Gallistel's classic monograph, *The Child's Understanding of Number* (1978). A.B. at three years six months was counting eight objects: *One, two, three, four, eight, eleben. No, try dat again. One two, three, four, five, ten eleben . . . One, two, three, four, five, six, seven eleben! Whew!*[13] To count objects, children must link each counting word with exactly one object to a counting word. Gelman and Gallistel call this the *one-to-one correspondence principle*. The words must always be in the same order – *the stable order principle*. A third principle I outlined in Chapter 1 is the *cardinal principle*: the last word of the count denotes the number of objects in the set of objects to be counted; for example, when counting the objects '*one, two, three*', three is the number of objects counted. So even if the word is wrong, as in *eleben*, the principle has been applied.

Now, reciting the counting words in the correct order is a pretty trivial achievement for most adults. When I was working as a neuropsychologist, we saw patients with profound neurological damage who could do this recitation and very little else. But are these words really necessary to be able to count? Is there actual evidence from humans that these words are necessary?

Counting words: do we need them to count?

A key figure in the history of the cognitive development of children, the Swiss psychologist Jean Piaget (1896–1980), believed that the acquisition of counting words plays little or no role in the development of the concept of (cardinal) number, the number of objects in a set.[17] The critical idea was how the children came to acquire the concept of 'conservation of number' – that is, how some operations on sets, like rearrangement of the objects, did not affect the numerosity of a set, while adding or taking away an object did. He describes examples of tasks with children around six years old who used counting words but were no better at establishing the numerical equivalence of two sets than those who didn't.

However, since Piaget's time, it has been claimed by many influential and distinguished investigators that without counting words, humans couldn't count exactly above four – the limit of 'subitizing' – the number of objects one can accurately enumerate in a glance without sequentially counting them. Beyond four, it is claimed, we only have concepts of approximate numbers, not exact numbers. When we count with counting words, five is exactly one more than four and exactly one less than six.

So how do we get from this 'approximate (or analogue) number system' approach I described in Chapter 1 to our familiar exact number concepts in which fourteen is exactly one more than thirteen and one less than fifteen? In this approach, our 'starter kit' for counting and calculations contains the small exact number system and the large approximate number system. To get exact representations of these larger numerosities, according to the Harvard psychologist Susan Carey, it is necessary to have learned the list of counting words which enables us to 'bootstrap' from the exact small number mental representations to the exact larger number representations.[18]

One critical difference between our approach, with an accumulator system built in to the starter kit, and the approximate (or analogue) number system, is that in the latter, numerical magnitudes are scaled logarithmically. This makes simple calculation, adding and subtracting, complicated: you need to be able to add and subtract the logs of the numbers. Our approach with a linear accumulator means that these calculations are straightforward. I will return to this difference in the final chapter, where I show with careful experimentation designed to distinguish these different approaches that the internal representation of numerical magnitude in both animals (in this case, birds) and humans is indeed linear.

A more extreme view about the starter kit has been taken by Rafael Núñez, a mathematician and cognitive scientist from the University of California in San Diego. He argues for a fully cultural basis for number and numerical abilities, rather like for snowboarding. Although of course there are 'biologically evolved preconditions' – for snowboarding, bipedal balance, depth perception and so on – the actual practice is a cultural artefact; analogously, there will be biologically evolved preconditions for learning about numbers. For snowboarding there needs to be a specialized cultural tool, a snowboard; for a sense of number, the cultural tools include the counting words or the familiar Hindu-Arabic symbols 1, 2, 3. You just have to learn to count.[19]

Neither approximate number position nor the 'snowboard' position has gone unchallenged. Even today, in our highly connected world, there are languages that have very few number words, or even a word for number, and the words that they do have are not used for counting and do not refer to exact numerosities. These languages survive in remote places such as Amazonia and Aboriginal Australia. A recent survey of 189 Aboriginal Australian languages representing 13 language families reported that 139 (74 per cent) have an upper numeral limit of only 'three' or 'four', and an additional 21 languages (11 per cent) have a limit of 'five'.[20] On the face of it, these languages and cultures provide prima facie evidence for the necessity of counting words for counting.

According to Núñez, speakers of these languages cannot have a concept of number, and according to the approximate number approach, they can only have an idea of numbers above four that are at best approximate, whereas our concept is of exact discrete steps between numbers – from four to five is one step, from fourteen to fifteen is one step.

The distinguished mathematician Abraham Seidenberg (1916–1988) argued in his paper 'The ritual origin of counting' that counting was actually *invented*.[21] Not only that, but it was invented in a single place, somewhere in the Middle East millennia ago. The fact that counting is so widely used, he notes, does not indicate its origin. And the fact that counting 'appears to be the simplest of mathematical devices' does not mean that it was easy to discover, or that many people or cultures invented it again and again.

Critically, he provides numerous examples that counting is widely associated with rituals and religious beliefs, giving numbers mythic properties. For Pythagoras, for example, odd numbers are male and even numbers are female. God is One. The animals, one male and one female, went into Noah's ark two by two.

Another clue to its ritual and religious origin are the many taboos on counting. In one African society it is considered extremely unlucky for a woman to count her children in case the evil spirits hear her and take one away by death. The Masai do not count people or animals, for a similar reason. Orthodox Jews require ten males for particular rituals but are not allowed to count them; instead, each male recites one word of a ten-word sentence. 'Don't count your chickens before they are hatched' is a familiar taboo.

One particular ritual associated with the creation myth requires pairs to appear in the ritual procession: male/female, king/queen, light/dark and so on. This, Seidenberg suggests, prompted two-counting, which is still practised in remote regions with restricted number vocabularies, such that the languages count one, two, two+one, two+two, two+two+one and so on. He cites languages in Australia, Amazonia and South Africa which count in this way. (See p. 52 for a well-documented example from the Australian language Garrwa.)

One comment interested me particularly. Most languages have base systems of five, ten or twenty, which is obviously linked to the number of digits on our hands and feet. However, we do not need to count our fingers or toes. We know how many there are. According to Seidenberg, we started to use these base systems in order to go beyond two-counting when the ritual became longer and more complex. This is a fascinating diversion, but Seidenberg did not have access to modern archaeology, neuroscience or anthropology, and may well have been content to talk about the origin not of counting, but of counting practices.

One way of challenging Núñez, Carey and Seidenberg is to ask whether children raised in cultures where the language does not contain counting words are able to count and carry out simple calculations.

The first report of such a language and culture comes from a sixteenth-century missionary Jean de Léry (1536–1613) in his *Histoire d'un voyage fait en la terre de Brésil, autrement dite Amerique* (1578). It was subsequently quoted by the English philosopher John Locke in 1690. In Locke's words the 'Tououpinambos' (Tupinambá, a tribe from the Brazilian rainforest) 'had no names for number above five'. But they 'can reckon beyond five . . . by showing their fingers, and the fingers of others who were present'.[22] So the Tupinambá indeed were able to count beyond five without the relevant counting words. Of course, this is not a properly controlled study, let alone peer-reviewed, and maybe de Léry got it wrong.

I checked de Léry's report with perhaps the greatest expert on Amazonian languages, Alexandra Aikhenvald of James Cook University, Australia. She writes,

In most Amazonian societies, counting was not a cultural practice: a counting routine was absent. Forms nowadays translated as 'one', 'two' and 'many' were not used for enumeration. 'One' would have meant '(be) alone', 'two' would refer to '(be) a pair', and 'three' would cover 'few, or many' . . . [But] since the principle which underlies counting is present, filling the gap is a rather trivial matter.[23]

This is why Spanish or Portuguese counting systems are quickly mastered and used by the Amazonian peoples, especially in situations where the use of money is important. I asked Alexandra specifically about the account by de Léry. She wrote to me that 'Chances are high that de Léry was spot on!'

One widely cited study claimed that speakers of Munduruku, one of the Tupi languages that lacked counting words, could only mentally represent numerosities above four approximately.[24] Is this true? De Léry, as noted, reported that Tupi language speakers could use their fingers to carry out exact enumeration. In the photos accompanying the published Munduruku study, an elderly man can be seen counting on his toes. Similarly, the American linguist Kenneth L. Hale (1934–2001) noted that Australian Aborigines whose language does not contain counting words, like Warlpiri, nevertheless pick up English counting words very quickly when they need them, for example, as stockmen who count cattle, and when money is involved,[25] suggesting that the counting principles are in the heads of the Warlpiri even without a handy way of expressing them.

If this is correct, as indeed I am arguing in this book, and humans do have an innate capacity to count, it is not clear why these languages should lack counting words. In some remote cultures there are counting practices without specialized counting words. For example, in the remote valleys of the New Guinea highlands many tribes do not have special words for counting, but instead use body-part names to stand for specific numerosities. For example, the Yupno count up to 33 on their body, starting with 1 on the left little finger, to 10 on their right thumb, 26 and 27 on left and right nostril, and 31, 32 and 33 on the left and right testicles and penis.[26] Women, according to the investigators, do not count in public.

The New Guinea groups have a gift-exchange culture. The number of pigs given has to be remembered, if a suitable reciprocation is to be made. To keep track of these transactions, 'without a name or mark to distinguish that precise collection, will hardly be kept from being a heap in confusion . . . Distinct names conduce to our well reckoning,' as John Locke noted in *An essay concerning human understanding* (1690). So the Yupno and other groups used body-part names to remember the number of pigs.

It is worth remembering that much of our own numerical vocabulary is derived from body-part names: the word 'digit' denotes both digits and fingers and toes. The words 'five' and 'fist' derive historically from the same root. And so on. And of course, it is not a coincidence that most languages use a base system corresponding to the number of fingers, or the vigisemal (base-twenty) system that corresponds to fingers plus toes, which is still found vestigially in French (*quatre-vingts*) and in biblical English (three score years and ten); there is a fully twenty-based system in Basque.

Although there is plenty of evidence that ancient Australians traded widely, and over long distances, they seem to have bartered face to face: I give you this, if you give me that. For this kind of transaction, no counting words are needed. Although sign language was widely used for inter-tribal communication, they don't seem to have signs for numbers either, and they don't seem to make tallies on sticks, bones or stones.[27]

In fact, in hunter-gatherer societies in Australia and South America, almost all single lexical terms have an upper limit at three. Beyond three, the languages use compounds, like 'two and one' for three.

Here is an example from the Australian language, Garrwa:

2	*kujarra*	
3	*kujarra-yalkunyi*	(2 and 1)
4	*kujarra-kujarra*	(2 and 2)
5	*kujarra baki kujarra yinamali*	(2 and 2 and 1)
6	*kujarra baki kujarra baki kujarra*	(2 and 2 and 2)[28]

Also, ancient Australians were not agriculturists,* but hunter-gatherers, and thus had no food surpluses to trade on a seasonal basis, again unlike the inhabitants of the Fertile Crescent or New Guinea. Patricia Epps and her colleagues documented both hunter-gatherers and mixed subsistence (a bit of agriculture) in South America and Australia, and found a link between both types and lack of counting words.[28] But the causal relation is speculation. It is still a mystery as to why Australian and many Amazonian languages lack the usual complement of counting words. Could it be that the earliest speakers of Proto-Pama-Nyungan (Australian) actually had these counting words but they disappeared through disuse? (See Chapter 3.)

---

* Bruce Pascoe in *Dark Emu* (Scribe, 2018) presents evidence that Aboriginal groups in pre-colonial and early colonial times were astute land managers who planted, irrigated, harvested and stored roots like yams, and grasses with edible seeds. However, there does not seem to be evidence that surpluses were traded at a distance.

Counting and calculation without
counting words

With these considerations in mind, we tested two groups of monolingual children speaking two Aboriginal languages which have no counting words. One group spoke Warlpiri and the other Anindilyakwa. (I have added a note at the end of the chapter about the languages.) Not only are there no counting words, but there are no traditional counting practices, no body-part counting, and no tallying. However, we were encouraged by Robert Dixon, whose extensive knowledge of Australian languages and cultures suggested that with culturally appropriate tests, these children could reveal evidence of both counting and arithmetic. As I noted earlier, Kenneth Hale (an expert on Warlpiri linguistics) commented that

> Aboriginal Australians have no difficulties in learning to use English numerals. The English counting system is almost instantly mastered by Warlpiris who enter into situations where the use of money is important . . . This is sufficient proof, if proof were needed, that there is nothing in the intellectual endowment of Aboriginal Australians which prevented the development of numeral systems in these languages.[25]

Bob Reeve, an expert in child development at Melbourne University, and I set out to test one particular aspect of this 'intellectual endowment', the ability to count sets and carry out arithmetical operations on them. Bob and I designed the tests, and we were lucky to have two excellent and committed research-ers, Delyth Lloyd for the Warlpiri children and Fiona Reynolds for the Anindilyakwa, who were willing to go into the field to carry out the research.[29]

The Warlpiri children were from a community in Willowra, a remote desert settlement 400 km from Alice Springs; the Anin-dilyakwa children were from Angurugu, a community on Groote Eylandt off the north coast of Arnhem Land. Back in 2002, when we started, these places were not connected to the internet, and communication depended on monthly phone calls to a local com-munity centre or school, where our researcher may or may not be at the time of the call. It was also quite difficult to recruit the chil-dren, since Aboriginal life is very busy, and involves a lot of hunting and gathering away from the settlement, often for weeks at a time.

In the basic set-up, the experimenter, a local bilingual helper and a subject would sit on the ground and the task was explained. To test counting ability we used two basic tasks. The experimenter would put out counters one by one on to her mat, then cover the mat and ask the subject to make her mat the same. We couldn't ask the child to put out the same number of counters since there was not a word for 'number' in their language. So the task had to be explained by showing. The second counting task was cross-modal matching. The experimenter would bang two sticks together and the child had to put out as many counters as the number of bangs. In this, the child could not use the visual pattern to make the match.

For exact calculation, we also borrowed a very nice task from American developmentalists Kelly Mix, Jane-Ellen Huttenlocher and Susan Levine. Using materials from the memory task, the experimenter placed one or more counters on her mat and, after four seconds, covered the mat. Next, the experimenter placed another counter beside her mat and, while the child watched, slid the additional counter or counters under the cover and on to her mat. Children were asked by the Aboriginal assistant to 'make your mat like hers'. We tested nine additions: 2 + 1, 3 + 1, 4 + 1, 1 + 2, 1 + 3, 1 + 4, 3 + 3, 4 + 2 and 5 + 3 (see Figure 1).

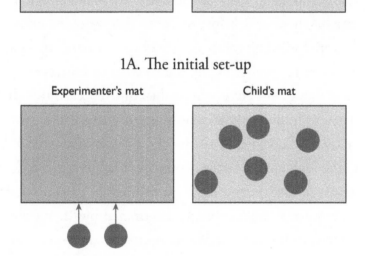

1A. The initial set-up

1B. Adding two

Figure 1. Non-verbal addition. 1A. The experimenter places four counters one by one on her mat, and then covers them. 1B. Two counters are added to the experimenter's mat in full view of the child so that they are underneath the cover. That is, the child cannot see the result of adding two to four and has to work it out, and then place the computed result on his or her own mat.

We compared the performance of the children in these two communities with English-speaking children of the same age in

Melbourne. In none of the tasks we tried were the Melbourne children better than the Aboriginal children. Not only did we fail to find a difference, but by doing the right statistics, we showed that groups were more likely to be the same than to be different. So we concluded that children without the advantage of counting words still have the concept of exact number and can carry out exact addition.

Fiona went back to Angurugu to carry out further tests on children there, including the exact calculation task. Here we were interested in whether for Aboriginal children there was a link between number and space as there is in our culture. Is it a prerequisite for them as it is for us? To do this we tested an ability called 'spatial working memory', which, as its name suggests, is our ability to remember briefly the location of objects in space. We know that in our culture individual differences in this ability are related to individual differences in arithmetical ability.[30] The test we used is known as 'Corsi blocks' after its originator, Philip Corsi. The test shows a board containing nine blocks. Blocks are tapped one at a time, and participants attempt to tap blocks in the same order. Memory span is determined by the number of blocks tapped without error. We also tested children on a 'culture-free' non-verbal test of intelligence known as Raven's Coloured Progressive Matrices. We wanted to make sure that any differences between the abilities of Anindilyakwa children and their counterparts in Melbourne were not due to this factor, though of course we knew that culture-free tests were in practice rarely culture-free. The Melbourne children were given the same sums but in their usual symbolic way: '2+1 = ?' etc.[31]

In both groups of children, the Corsi test was a good

predictor of individual differences, suggesting that the link between number and space applies also in non-numerate cultures. But there was one result that surprised me. There was a significant difference between the IQs of the Anindilyakwa children and the children in Melbourne: the former were fifteen points better than the latter. This wasn't the object of the study, so we didn't comment on it in our report, but there seemed to me two possibilities. The first is that the test isn't really a test of intelligence, but rather a test of spatial ability since it involves matrices, and indeed we found a correlation between the Corsi test and Ravens, and it has been known for many years that Aboriginal children and adults score higher on tests of spatial ability than their non-Aboriginal counterparts in Australia.[32] The second is that they really are at least as smart or even smarter. Alfred Russel Wallace (1823–1913), the co-originator of the theory of evolution, who spent many years in remote parts of the world with local tribes, wrote, 'The more I see of uncivilised people, the better I think of human nature on the whole, and the essential differences between civilised and savage man seem to disappear.' Indeed, Jared Diamond has suggested that New Guinea hunter-gatherers have to live on their wits, and so the main driver of reproductive success is intelligence, whereas the resistance to pandemic diseases that originate in our close contact with animals we breed is the main driver in settled agricultural or industrialized societies.[33]

This study shows that even without counting words and with no traditional cultural practices, children can count and can carry out simple calculations at least as well as their English-speaking counterparts growing up in highly numerate Melbourne.

Very young counters

If I am right that children grow up with the same cognitive capacities for number irrespective of their cultural and linguistic circumstances, then it should be possible to detect this capacity very early in life, perhaps even in the earliest infancy.

In Chapter 1, I described a useful way of testing whether an individual – human or other animal – can represent the numerosity of a set. It's called 'match-to-sample'. It is rarely used with humans, probably because it is easier to ask them the number.

Here is a very nice example of match-to-sample from Elizabeth Brannon's lab at Duke University.[34] Children aged three to four were tested to see if they could match a sample number of objects to one of two choices. The colour, shapes and sizes were varied so that the correct choices could only be made by picking the one with the same numerosity (see Figure 2). Of course, some of these young children may already know how to use counting words in a way that could help with this task.

Infants can count

The implication of our studies of Aboriginal children is that even without cultural counting practices, they are still able to count and carry out some arithmetical procedures. This suggests, and we will see much more evidence for it, that we humans inherit a counting mechanism which, in my view, is the accumulator system I described in Chapter 1.

If this is the case, then we may be able to see it in humans before they have had a chance, or much of a chance, to acquire

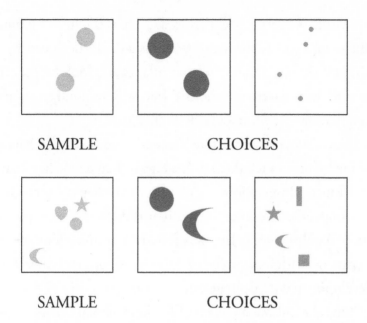

SAMPLE CHOICES

SAMPLE CHOICES

Figure 2. In this experiment, children aged three to four were shown a sample panel on a computer screen. When they had touched the sample, two new panels would appear and they had to choose the one that matched in numerosity. The study varied colour, shapes and sizes.[34]

anything from the culture. There is now forty years of research showing that infants, even in the first weeks of life, are sensitive to numerical changes in the environment. Of course, we can't obtain verbal reports from these subjects, nor can we set them numerical tasks. What we can do is observe how they respond to numerosities. For this, we have to be sure that it really is numerosities that they are responding to, not some other feature of the environment. This means we have to use a special methodology to investigate the numerical abilities of creatures that cannot use language. One way of demonstrating this is to see if an infant notices changes in the number of objects in a display.

Of course, when the number of dolls, for example, is increased, other visual properties are increased, such as the total amount of dollness on display, the area of the visual field occupied by the dolls, and so on. It turns out that infants will look longer when one doll is replaced by another but will respond even more strongly when the number of dolls changes.

This satisfies part of Gallistel's second condition (see Chapter 1) that the infant can tell that set A is bigger than or smaller than set B (arithmetical operations < or >). But can they carry out other mental operations isomorphic with arithmetic, for example addition and subtraction? This was tested in an ingenious series of experiments with four- to five-month-old infants by Karen Wynn, a developmental psychologist then at the University of Arizona.

What she did was a variant of Rochel Gelman's 'magic experiments'.[13] If the child has an arithmetical expectation, it can be tested by a 'magic' manoeuvre that produces an unexpected outcome. In this study, there was a small stage and a screen that could conceal it. One or more dolls were shown on the stage, and then the screen came down so that the doll/s were no longer visible. A big hand was shown putting another doll behind the screen. Now remember that the infant cannot see what's on the stage, but can only remember what was there before the screen came down and what happened when another doll went behind the screen. So any arithmetical work is done in the infant's head. Does the infant now have an arithmetical expectation about the result of the addition? If so, looking time should differ between the arithmetically expected outcome and the unexpected outcome. Wynn used three conditions: $1 + 1 = 2$, $1 + 1 = 1$, $1 + 1 = 3$. In the unexpected outcomes where the number of dolls 'magically' becomes one more or one less than expected, the infant looked significantly longer than the expected outcomes.

She also tested the infant's ability to carry out a mental operation isomorphic with subtraction. Here she showed two dolls; then the screen came down and one doll was removed. When the screen was raised, the infant saw either correct subtraction, one doll, or the incorrect subtraction, two dolls. The infant looked longer at the incorrect subtraction.[35]

Wynn, now at Yale University, and her colleague Koleen McCrink followed this up with older infants, aged nine months, and larger numbers, five and ten. Could they still carry out mental operations equivalent to addition and subtraction? This time, instead of dolls, stages, screens and hands, computer screens showed the rectangles and bars. The addition situation was this: the infant saw five objects which were then completely obscured by an occluder. Five more objects were seen to move behind the occluder. The occluder was removed to reveal either five objects for half the trials or ten objects for the other half. The infants looked longer at the arithmetically unexpected outcome, five, than the expected outcome, ten. Again, remember that the infant has to be processing the numbers in his or her head before the occluder is removed.[36]

For subtraction, the infant sees ten objects on the screen which are then obscured by the occluder. Five objects are then seen to be removed from behind the occluder, so that the arithmetic is 10–5. When the occluder is removed, what is left on the screen is either ten objects or five objects. The infant now looks longer at ten objects than five objects: the correct arithmetical result. Of course, Wynn and McCrink did all the appropriate controls about which arrangement an infant might naturally prefer and whether they were tracking the total amount of object stuff on the screen.[36]

Wynn and McCrink conclude that these results 'support the existence of a magnitude-based estimation system (such as the accumulator mechanism proposed by Meck & Church, 1983 [see p. 19]). Several researchers have proposed that this same mechanism underlies infants' numerical abilities.'

By six months, infants' number sense is already quite abstract in that they will match an *auditory* display of three voices with a visual display: three faces, not two faces, and two voices to two faces, not three.[37] In fact, even in the first week of life, infants will notice matches between the number of sounds and the number of visible objects on display.[38]

So it is not outlandish to propose that infants inherit a mechanism that enables them, without instruction, language or learning, to respond to the number of objects in their environment and to carry out arithmetical operations on them.

Seeing the world in numbers

According to David Burr, an Australian visual scientist working in Pisa, Italy, we have a visual sense of number. That is, we see numerosities in the world just like we see colour. It is automatic and affects our behaviour even when we do not attend to it. Burr and his colleague John Ross produced the first demonstration of that our visual system automatically 'adapts' to numerosity, that is, how many objects you think you see depends in part on the number of objects you have just seen. This is just like adaptation to movement, as in the waterfall illusion: when you look away from the waterfall – objects moving in one direction – then stationary objects seem to be moving in the opposite direction. Burr and Ross show that 'perceived numerosity is susceptible to adaptation', like movement. 'Apparent numerosity was decreased by adaptation to large numbers of dots and increased by adaptation to small numbers, the effect depending entirely on the numerosity of the adaptor'.[39]

I would add that while most of us see the world in colour, some of us may not see the world in numerosities. This becomes especially important when we think about the starter kit needed for learning about numbers and arithmetic. See p. 76 on heritability.

The starter kit

I argue that a counting mechanism, such as an accumulator, is an essential component in the human *starter kit* for learning arithmetic in school, at home or in other ways. How might it work?

Perhaps the first thing the child in a numerate culture has to learn is the meaning of the counting words. As I have said, this is actually more complicated than it sounds and may take several years to master completely. Just learning the sequence of counting words can occur in stages, as the example from three-year-old A.B. earlier shows. There is the stage at which the child recognizes the individual counting words – *one, two, three* as contrasted with the word *onetwothree*. Then there may be a stage at which the child realizes that the words must be in a stable order (the 'stable order principle') without necessarily knowing all the words up to, say, ten or even getting the words known into their usual adult order. Finally, there is the stage at which all the words to *ten* are known and in the correct order.

Of course, the child must learn how to use the words to count objects based on the principles of one-to-one correspondence between the words and the objects in the set, plus the order-irrelevance, abstraction and cardinality principles. This means recognizing from parents or carers contexts for counting.

But how does the accumulator relate to acquiring these cultural practices? The accumulator memory will be linearly proportional to the number of objects counted. If you think of it as a column, the higher the column the more objects are counted. The problem then is to link each counting word to a specific height of the column (see Figure 3). Now, it may take a while to calibrate the column and make sure that each word is linked to the appropriate height. Of course, the brain is noisy, and so the column height won't be exact. This is a well-understood effect that is seen in all kinds of neural function, and it's called 'scalar variability': that is, noise (variability) will be a function of the final count.

The tally-counter (Chapter 1, Figure 1) is a *linear accumulator* with a memory. That our brains possess such a mechanism is an old idea that actually comes from work with animals.

The linear accumulator has several attractive features. It meets Gelman and Gallistel's three counting principles described earlier: the *abstraction principle* – anything the selector can identify can be counted; the *order-irrelevance principle* – it doesn't matter which object is counted first, second and so on; and the *cardinality principle* – the last number in the count denotes the numerosity of the set, which is equivalent to the final level of the accumulator is the total of the count.

An ingenious experiment by John Whalen of Delaware University with Gelman and Gallistel shows that under the right conditions adult humans count in exactly the way predicted by the accumulator model. Here's the task they used, which, quite deliberately is based on a method used with rodents, as we will see

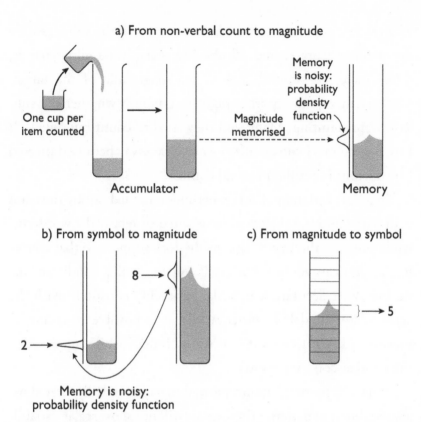

a) From non-verbal count to magnitude

One cup per
item counted

Accumulator

Magnitude
memorised

Memory
is noisy:
probability
density
function

Memory

b) From symbol to magnitude

8 →

2 →

Memory is noisy:
probability density function

c) From magnitude to symbol

} → 5

Figure 3. A model for learning the counting words based on the accumulator model (see Chapter 1). Each object or event counted increases the contents of the accumulator by a fixed amount, here two activation units. The nature of the object or event does not affect the number of units. There is noise in the system that is proportional to the number magnitude – larger numbers are noisier, indicated by the slope in the accumulator. This is called 'scalar variability'. The child learns to link the counting words to levels in the accumulator.[40]

in Chapter 5. Participants were required to press a key a specific number of times as fast as possible. Human adults can press a key about once every 120 msecs, while verbal counting rate is about half that, 240 msecs per word. In one condition, they were given a target number between seven and twenty-five and had to press a key as fast as possible to reach that target. All the subjects were reasonably accurate, and all showed scalar variability, that is, errors increased proportional to the magnitude of the target. Many analyses and reports sought to establish whether the subjects were counting. Many said they tried to count but failed: 'If I tried to count, I consistently lost track anyway because I pressed [the button] faster than I could count.'

In a neat variation of this experiment, carried out by the team with Sara Cordes, subjects were required to repeat the word 'the' while pressing the key.[41] This really does suppress verbal counting, as you can verify for yourself. Again, in this condition, the subjects were accurate with scalar variability consistent with the accumulator model. In other words, it is possible to count by pressing a key n times without using, indeed while suppressing, the familiar counting words.

In fact, if participants are allowed to count verbally, they show a quite different pattern. Their errors will not show scalar variability. If you count slowly, you will most likely arrive at the correct count. Errors will occur if an object is missed or double counted. This means that variability will be binomial rather than scalar – that is, the variability increases more slowly with number size. And this is exactly what Cordes and the team found.

To infinity and beyond?

Humans, even quite young humans, can count up to pretty high numbers. For example, in one study of ethnically and economically mixed kindergarten children in Pennsylvania by Patrice Hartnett and Rochel Gelman, half were able to count accurately from 101 to 125, and they then probed if the child could count higher.[42] However, the purpose of this study was not to see how high children could count but to see if children of this age and Grades 1 and 2 – five-to-eight year olds – had a concept of infinity.

In Chapter 1, I described the position taken by John Locke who said that the simplest idea is *one*; *one* can be repeated; 'by adding one to one, we have the complex *idea* of a couple', etc. He argued that numbers give us the clearest idea of infinity, which 'lies only in a power still of adding any combination of units to any former number, and that as long and as much as one will; the like also being in the infinity of space and duration, which power leaves always to the mind room for endless additions.'[22]

Do people, even children, really think like this? Do they really think it is possible to go on adding one with no end? This is what Hartnett and Gelman wanted to discover. They explored this by asking the children questions like the following:

- Can people always add to make a bigger number, or is there a number so big we couldn't make it any bigger?
- If we count and count, will we ever get to the end of the numbers?
- What if we cheat and instead of starting at one, we start counting from a really high number? Then could we get to the end of the numbers?
- Is there a last number?

Here's an example of a seven-year-old who they classified as an 'Understander'.

If I thought of a really big number, could I always add to it and get a bigger number? Or is there a number so big that I couldn't add anymore; I would have to stop? *You could always make it bigger and add numbers to it . . .* If I count and count and count, will I ever get to the end of the numbers? *Uh uh [No].* Why not? *Because there isn't one.* There is no end to the numbers? *Uh uh.* How can there be no end to the numbers? *Because you see people making up numbers. You can keep making them, and it would get higher and higher . . . you can just keep on making up letters and adding one to it.*

Here's an example of a six-year-old 'Non-understander':

Why do we have to stop [counting]? *Cause you need to eat breakfast and dinner.* Yes. But then you could start it up again after you ate. *You forget where you stopped . . .* What if we got to that number [this child's biggest number], and we tried to add one anyway? What would happen? *I guess you'll be old, very old . . .* How will you know when you get to the end? *Well, you can stop any time you want to.*

There were also 'Waverers', who gave inconsistent answers.

It was found that children able to count accurately beyond 100 were more likely to be among the roughly 40 per cent of all the children sampled who were classified as Understanders. Up to 67 per cent of Grade 2 children, and 15 per cent of kindergarten children, were classed as Understanders. So maybe infinity is not such a difficult concept after all.

It may be that children who have mastered the counting word list infer that there is a linguistic procedure for making up larger and larger numbers, and that certainly fits with the finding that better counters are more likely to be Understanders. Alternatively, the children may have a sense about how the counting words represent sets of objects. A recent study by David Barner and his colleagues found that children who understand the cardinal principle – that the last count word is the numerosity of the set counted – are more likely to be Understanders.[43] But I wonder if you really do need the cardinal principle.

One of my favourite experiments in this area is by Barbara Sarnecka and Susan Gelman, which suggests a somewhat deeper explanation even though it was with much younger children, those aged between two years five months and three years six months.[44] Here's the set-up. The child observes the experimenter counting aloud 'one, two, three, four, five, six' as she puts 'moons' into a box one at a time. She then checks that the child has understood and remembered what has been done by asking 'How many moons in the box?' The child must answer 'six' to establish that he or she had been paying attention and remembered the last word of the count, before going on to the next part of the experiment. Now the box, which is closed, is shaken vigorously, and the child is asked the question 'How many moons in the box?' Overwhelmingly the child says 'six'. In a second condition, a 'moon' was added or taken away from the box in full view of the child. Again the question was how many moons are in the box. Overwhelmingly the child gave a number word that wasn't six; it could be a number word referring to a number larger or smaller.

Here's the really interesting part: these very young children were unable to correctly give all the numbers up to six objects in a give-a-number task, for example, 'Can you give five apples to the monkey? Just take five and put them right here on the table in front of him.'

Children were then grouped according to the highest number

word whose exact meaning they knew. Most of them could do this accurately for one and two, but not for many higher numbers. Even those who were correct for six often failed with lower numbers. But crucially, what these children did know was that number words referred to a specific number of objects even when they didn't know what exactly the counting words meant. When the number was changed by adding or subtracting an object from the set in the box, they knew that the number word had to change. When the set was rearranged by shaking the box, they knew that the number was, to use Piaget's term, 'conserved'. Adding one, as Locke long ago noted, will be recognized as changing the number even though you don't know what name to call that number.

In terms of the theory set out in Chapter 1, children understand that sets have a specific numerosity and that certain operations on these sets ('isomorphic with arithmetical operations') can change the numerosity, while others which are not isomorphic with arithmetical operations (such as rearranging by shaking the box in this study) do not change the numerosity.

Is there an accumulator in the human brain?

My claim is that the accumulator is a very simple mechanism, though the selector is not. Now, the human brain contains over 85 *billion* neurons (brain cells) and trillions of connections among neurons, so looking for a tiny mechanism will be rather like looking for a very tiny needle in a very big haystack.

Our current methods of finding needles in the haystack are neuroimaging and the effects of brain damage, and these will at best tell you roughly where the needle may be – somewhere in the top left of the haystack, for example. This is because neuroimaging methods currently available measure millimeter cubes of brains, 'voxels', each of which contains over half a million neurons, 2 million glial cells, and an enormous number of connections. And neuroimaging studies cannot identify single voxel activity but only the activity of a bunch of voxels that are related to the cognitive function of interest. Brain damage, often caused by big changes to blood flow (strokes) or tumours, will involve hundreds of voxels. Nevertheless, we have discovered some useful evidence.

First, we have known for a hundred years that there was a special brain network for number processing of all kinds. This was due to Salomon Henschen (1847–1930), a Swedish neurologist, and the first person to use the term 'acalculia' (*akalkulia*, in German). He discovered from his own patients, and those described in the literature, that acalculia was a selective deficit in numerical abilities that was quite independent of language abilities. He identified the left parietal lobe (we'll hear much more about the parietal lobe in Chapter 4 on primates). He also found that there were, in the adult brain, subcomponents of mathematical ability separable in the brain. First between input processes and output, motor, processes. He also distinguished number words, *Zahlen*, from digits, *Ziffern*. Henschen cited 122 cases of 'word blindness' (alexia) in which seventy-one of these subjects could still read digits, while fifty-one could read neither. These cases were the result of damage to the left parietal lobe. He also observed that damage to what we now call Broca's area, the third left frontal convolution, would impair verbal counting. His original observations have been extensively supported by more recent very detailed studies of individual patients with and without damage to the parietal lobes.

One good example is Signora G, a patient of the Paduan

neurologist Franco Denes studied by my then-student, Lisa Cipolotti. Signora G suffered a stroke that left her with extensive left hemisphere damage including to the parietal lobe. Before her stroke, she kept the books of her family's hotel, and was arithmetically very competent. Now, alas, she was unable to count verbally beyond four. So when Lisa asked her to count five objects, she correctly counted four but when she got to the fifth, she said, *La mia matematica finisce qui* ('My mathematics finishes here'). She couldn't select the larger of two digits or the larger of two numerosities if they were larger than four. When the number of objects was in the subitizing range, up to four, she would count them rather than immediately recognize the numerosity. At the same time, her reasoning, knowledge of non-numerical symbols (such as the logos of the many Italian political parties) and memory were unimpaired.[45] Somewhere in the left hemisphere, as Henschen discovered, her counting system was damaged.

Elizabeth Warrington, one of the founders of modern cognitive neuropsychology, at the National Hospital for Neurology and Neurosurgery in London, with her colleague Merle James, collected a case series of patients with left or right brain damage. In one task, she asked them to estimate the number of dots and dashes from three to seven presented on cards very briefly (100 msecs) in a tachistoscope (those were the days before readily available computers). They discovered that patients with right parietal damage were very impaired in estimating the numerosity when the dots were presented to the contralateral visual field – that is, to the left visual field – because this projects mainly into the right hemisphere. And this wasn't because they were missing out some of the dots, since most errors were of overestimating the number.[46] Those with left parietal damage were not significantly worse than controls.

In a neuroimaging study, we also found a right hemisphere region involved in small numerosity enumeration.[47] However, almost all neuroimaging studies of numerical processing in more or less whatever task activate both left and right parietal lobes.[48]

As I have mentioned, the accumulator should be able to count anything the selector selects. The number of legs on a table is four, whether you see them all at once or count them one at a time. My colleagues Fulvia Castelli and Daniel Glaser and I designed an imaging experiment using fMRI (functional magnetic resonance imaging) to see if this was indeed the case.[49] Instead of table legs, we used blue and green squares presented all at once in an array or in a separate condition, one at a time. The task was a very simple one: is there more green or more blue? We didn't even mention numerosities because we wanted to find out if the brain responded automatically in the two conditions: simultaneous and sequential presentation. We also had a control condition with equivalent quantities of blueness and greenness, but no separate objects because we wanted to be sure that the brain activation measured was specific to sets of blue and green objects, not just the total quantity of blueness and greenness. We took the activations measured when in these continuous quantity conditions, and subtracted them from the conditions where there were the separate objects. Of course, if the brain responded just to the quantities of blue and green, then the subtraction would have yielded nothing. However, there was a distinctive pattern of activation for the separate objects and there were small regions in the left and the right parietal lobe called the *intraparietal sulcus* that were active in both conditions. What is more, activations were higher the more difficult the comparison, so it was more active for eleven green squares compared to nine blue than fifteen green squares to five blue; in fact, activation increased monotonically with ratio of blue and green squares (see the section on Weber's Law in Chapter 1). A key characteristic of the numerosity of a set is that it is abstract. It should not matter how the objects are presented, all at once or one at a time, provided of course that the selector can select them. Here we showed that the same region responded to numerosities irrespective of the mode of presentation. See Figure 4 (p. 76).

I also wanted to know if the same brain region was involved in counting objects whether they were presented visually or auditorily. If so, this would be evidence for the same counter being used for both modalities. In this study, with my student Manuela Piazza, we used fMRI to measure the brain responses to a sequence of red and green squares where the subject had to say at the end of the sequence whether there were more red or more green squares. We also had the participant count the number of switches from red to green. The same participants also listened to a sequence of high and low tones, with the same temporal pattern as the sequence of squares, and this time to decide whether there were more high or more low tones, and also to count the number of switches. The same small region, the *left intraparietal sulcus*, is activated whether the stimuli are presented visually or auditorily.[50]

The human brain, therefore, does treat numbers as abstract – as properties of sets – so as well as being independent of mode of presentation – sequential or simultaneous – the same small brain region responds whether the objects are auditory or visual.

The problem with these findings is that only left hemisphere damage affects counting. So it remains a mystery what the right parietal is doing, at least in the numerate adult brain. Does it act as a kind of back-up? Does it just carry out very basic tasks, while the left does both basic tasks and more complex ones?

If adult counting is based on an accumulator mechanism, where is it located in the brain? There is a very clever study from Seppe Santens, Chantal Roggeman, Wim Fias and Tom Verguts at Ghent University in Belgium that shows that something like an accumulator, which they call 'summation coding', is located in the posterior superior parietal cortex close to the intraparietal sulcus in the left and right hemispheres.[51] The response of small regions became stronger according to the number of dots in the range one to five. That is, the larger the numerosities, the greater the activation in just these regions. In their experiment, the participants were required to attend to the numerosity of dots by asking them which digit matched the number of dots they had just seen on 12 per cent of occasions. The rest of the time they were not asked. This meant that the brain was automatically responding to the number of dots in the display. Of course, they carefully, and rather elaborately, controlled for the non-numerical parameters of the dot displays. Now, the posterior superior parietal cortex is equivalent to a region in monkey brains called the lateral intraparietal area, and this carries out a similar accumulating role, as I explain in Chapter 4.

There are few studies of the prodigious calculators mentioned earlier. It turns out their exceptional abilities are based on the same brain network as you and I use. In a study of Rüdiger Gamm's brain, Mauro Pesenti of Université Catholique de Louvain, Belgium, and colleagues found that his calculation processes recruited the same neural network as previously observed for both simple and complex calculation, *plus* a system of brain areas to extend working memory.[53] More generally, the brains of professional mathematicians use the same fronto-parietal network, and especially the parietal lobes.[54]

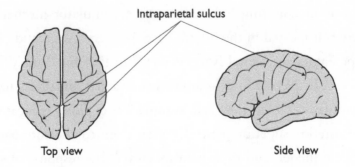

Figure 4. The intraparietal sulcus showing the two regions that are almost always active when the human brain is carrying out a numerical task. 4A is looking down on the brain at about the level of the ears. 4B is a side view of the left hemisphere.[49]

I have focused on sighted humans, but the brains of the early blind activate the same fronto-parietal network in numerical tasks,[48] supporting the idea that humans inherit specialized neural structures for number processing that do not depend on learning from seeing sets of objects.

Of course, adult human numerical abilities, especially in numerate cultures, are extensive and varied, as I mentioned earlier, and many brain areas are involved, but the core is in the parietal lobes.

Heritability

Are these numerical capacities part of our inherited cognitive endowment? I argued in *The Mathematical Brain* (1999) that humans inherit a 'number module'. By the word 'module' I meant that it conformed to the philosopher Jerry Fodor's influential definition. It was innate, domain-specific and applied just to numbers; its processing was mandatory and automatic – that is, you couldn't help seeing numerosities in the environment; it was implemented in a specialized neural apparatus. This was way before Burr and Ross's experiment. There was only one study of twins looking at the inheritance of numerical abilities,[55] and there was very little, and rather primitive, neuroimaging. Infant studies were, shall we say, in their infancy. At that time, I was a bit vague about how the module worked. Now I would say, as I argue in this book, that the central mechanism in the human number module is the accumulator system located somewhere in the parietal lobes of humans.

So, do we inherit a number module from our parents, as I claimed back then?

One way of investigating this is by looking at twins. Identical twins (monozygotes) have identical genes, while fraternal twins (dizygotes) share on average only half of each other's genes. So to assess the heritability of a trait, for example numerical ability, identical twin pairs should be more similar to each other ('concordant') than fraternal twin pairs. A very large study of seven-year-olds by Yulia Kovas at Goldsmiths, University of London and colleagues found that a third of genetic variance in seven-year-olds is specific to mathematics.[56] A later study by Kovas and her team found that one measure of basic numerosity processing, dot comparison ('Are there more yellow or more blue dots?'), is moderately heritable in sixteen-year-olds.[57]

If this basic numerosity-processing capacity is heritable, then in some cases, there could be a genetic anomaly in which inheritance fails.

Think about colour vision. We inherit the capacity for seeing the world in colour, except that some of us don't. This has nothing to do with our intelligence, memory or social background. It is usually inherited, though it can very occasionally be due to diseases of the retina or the optic nerve.*

Is there any equivalent for numbers – a kind of 'number blindness' – and if so, is it inherited?

First, it is not clear that a numerical disability is exactly analogous to colour blindness where the colour is not perceived or distinguished from other colours – green from red or blue from yellow. It's more like dyslexia, sometimes called 'word blindness', except that the dyslexic can actually see the word, but they have trouble interpreting it or pronouncing it, though a minority of dyslexic readers do have trouble seeing the words in the normal way.[58]

About 5 per cent of people are born with a condition called 'developmental dyscalculia'. These individuals are very poor at estimating the number of objects – either very slow, because they are counting one object at a time, or much more error-prone in counting than their peers. They are also poorer at selecting the set with more objects from two sets: their *Weber fraction* is significantly larger – that is, they need a bigger proportional difference between the sets to make the correct selection reliably.[59]

So one clue to heritability is if developmental dyscalculia is inherited.

There is one early twin study of very poor numerical ability in forty identical and twenty-three same-sex fraternal – non-identical – twin pairs where at least one twin had an arithmetic

---

* Red/green colour blindness (a colour vision defect) is due to a change in a gene (OPN1LW or OPN1MW) passed from mother to son on the twenty-third, X, chromosome – the sex chromosome. Because males, XY, have only one X chromosome, and females have two X chromosomes with a copy on each, XX, this kind of colour blindness is more frequent in males than females. Blue/yellow colour blindness is inherited in a different way, which means one copy of a different altered gene (OPN1SW) is sufficient to cause the condition.

difficulty. It was found that the identical pairs were more likely to be concordant, that is, both had arithmetic difficulty. Now, they weren't directly testing for something like number blindness but rather arithmetic.[55]

We have recently studied a sample of 104 monozygotes, and 56 dizygotes with a mean age of 11.7 years. There was greater concordance between monozygotic twin pairs than dizygotic twin pairs in their dot enumeration efficiency and in abnormalities (reduced grey matter density) in the critical region of the left parietal lobe.[60] However, anything that affects the development of the parietal lobes will be a factor, and several non-heritable factors can cause this: foetal alcohol syndrome, where the mother drinks too much alcohol during pregnancy, is one.[61] Very low birthweight can affect both the critical region in the parietal lobes and numerical processing.[62]

Another cause is Turner syndrome, which affects one X chromosome in females but isn't normally inherited since affected females are usually infertile. It is just bad luck. This condition is known to affect the parietal lobe and also very basic numerical processing in almost every Turner syndrome female appropriately tested and reported.[63]

Another X chromosome condition that is inherited is Fragile X syndrome. In females, the relevant gene is called FMR1. One region of the FMR1 gene contains a particular DNA segment known as a 'CGG trinucleotide repeat', so called because this segment of three DNA building blocks (nucleotides) is repeated multiple times within the gene. In most people, the number of CGG repeats ranges from about five to about fifty-five, but if the number goes above fifty-five, there can be symptoms, and above 200 repeats, you get the very serious Fragile X syndrome.

One revealing study by my colleague from Padua University, Carlo Semenza, and his colleagues explored females with more than fifty-five CGG repeats but who didn't show Fragile X symptoms. The eighteen female subjects, all with normal intelligence, undertook a very extensive battery of mathematical tests, from basic numerical tasks to complex arithmetic. It turned out they were significantly worse than matched controls on very simple number tasks such as counting dots and number comprehension where they had to choose the larger of two multi-digit numbers, and where they had to choose from among three alternatives the position on a number line from 0 to 100 of two-digit numbers. However, on more complex calculations, they performed like the controls, suggesting that there is something in the X chromosome that is crucial for building simple basic numerical capabilities.[64]

Does this mean that the gene or genes for numerical ability are only found in the X chromosome? One genome-wide association study (GWAS) may have found a variant in Chromosome 3. A German GWAS found a variant on a gene in Chromosome 22 associated with poor numerical abilities, along with abnormalities in the intraparietal sulcus. That looked promising, but another very large GWAS, led by Silvia Pararcchini at St Andrews University in Scotland, failed to replicate this finding with English cohorts. So we're still gene hunting. My current approach is to start with animal models, where the genome can be manipulated to test which genes are critically linked to dyscalculia. See Chapter 8 on fish![65]

So we have seen that humans really can count, and do things with the results of counting – some humans to an extraordinarily high level. Although to achieve this level requires training and commitment, we have also seen that the basis of this ability is not learned, since it can be observed in cultures without counting practices or counting words, and can be detected even in infant humans. There is a specialized brain network that supports counting and calculation in the parietal lobes of the human brain that seems to house one, or perhaps many, accumulators. While we are born able to count, the genetic basis of its inheritance is not yet established. Like colour blindness, some people seem to be born with a serious difficulty in carrying out even simple calculation tasks.

Note on Warlpiri and Anindilyakwa languages[29]

Warlpiri is in the Pama-Nyungan language family. It is a classifier language with three generic types of number words: singular (*jinta*), dual plural (*-jarra, jirrama*), and greater than dual plural (*jirrama manu jinta, marnkurrpa, wirrkardu, panu*). Anindilyakwa, probably unrelated to any other Australian language, is the major indigenous language spoken on Groote Eylandt. (It is also spoken in some small communities on neighbouring islands and on the nearby East Arnhem Land coast.) Like Warlpiri, Anindilyakwa is a classifier language, with nine noun classes and four possible number categories: singular, dual, trial (which may in practice include four) and plural (more than three). Anindilyakwa has a base-five number system, apparently appropriated from the Macassan traders who visited the northern coast of Australia, including Groote Eylandt, from about the seventeenth century onwards. It appears to be the case that the base-five system is reserved for special cultural enumeration events (such as distributing turtle eggs to recipients). In Anindilyakwa, numerals are adjectival, and must agree with the nouns they qualify. Because there are nine noun classes, enumerating in Anindilyakwa is complex.

However, the number names are 1 (*awilyaba*), 2 (*ambilyuma* or *ambambuwa*), 3 (*abiyakarbiya*), 4 (*abiyarbuwa*), 5 (*amangbala*), 10 (*ememberrkwa*), 15 (*amaburrkwakbala*) and 20 (*wurrakiriyab-ulangwa*). The word for 20 is invariable – that is, it does not change its form in different grammatical contexts. The Anindilyakwan number system is not formally introduced to members of the community until they reach adolescence. Judith Stokes, a pioneering documenter of Anindilyakwa and Groote life, observed that 'In traditional Aboriginal society nothing used to be counted that was outside normal everyday experience. When asked for what purpose counting was used in the old days, *the old women who know the number names* [emphasis added] say that counting was used for turtle eggs.' Although these languages contain quantifiers such as 'few', 'many', 'a lot', 'several', and so forth, these are not relevant number words because they do not refer to exact numbers. Ordinals, such as 'first', 'second', and 'third', would be more problematic. However, these words do not exist in either Warlpiri or Anindilyakwa.

CHAPTER 3

BONES, STONES AND THE EARLIEST COUNTING WORDS

We can find evidence from the earliest historical records, six thousand years ago, that humans were already counting and were able to carry out quite complex calculations. In prehistory, before there was writing, Stone Age humans more than ten thousand years ago were counting on bones, stones and cave walls, and also using counting words. In Chapter 2 I showed that Aboriginal Australian children who spoke only languages without counting words when asked to carry out numerical tasks that didn't require these words performed as well as English-speaking children. On the basis of this evidence, and from the behaviour of pre-linguistic infants, I argued that we are born with a counting mechanism in our brains, the 'accumulator'. This suggests that even without writing, humans, and perhaps other members of the genus *homo*, were able to count and carry out calculations based on counting.

The question I ask in this chapter is: when did genus *homo* first invent a counting technology, and why? These inventions are not confined to *Homo sapiens* – anatomically modern humans – but there is evidence for counting in other *homo* species, such as *Homo neanderthalensis* (sixty thousand years ago), or even more anciently, *Homo erectus* (about 2 million years ago).

Counting in history

In remote historical times, when people already possessed writing systems, our ancestors recorded numbers, and carried out calculations – sometimes quite sophisticated calculations.

For example, we know from the 6-metre-long Rhind Mathematical Papyrus in the British Museum that ancient Egyptians around 1550 BCE had manuals to teach them how to count and calculate. As Neil McGregor, Director of the British Museum, explained,

> If you wanted to play any serious part in the Egyptian state, you had to be numerate. A society as complex as this needed people who could supervise building works, organise payments, manage food supplies, plan troop movements, compute the flood levels of the Nile – and much, much more. To be a scribe, a member of the civil service of the pharaohs, you had to demonstrate your mathematical competence. As one contemporary writer put it, 'So that you may open treasuries and granaries, so that you may take delivery from one corn-bearing ship at the entrance to the granary, so that on feast days you may measure out the gods' offerings' [Papyrus Lausing BM 9994].

The papyrus was written in a hieratic script, easier to write than hieroglyphics, suggesting that this was not a sacred document, but a practical one. Unfortunately, few if any Egyptian mathematical papyri survive.

Earlier, Babylonians made marks in clay, *cuneiform*, which have survived much better than fragile papyrus. They used only two symbols, ⊤ for one and ⟨ for ten. ⊤ could mean one, or in a different position one times the base. ⊤ ⊤ would mean two times the base, and so on. Like our numbers, Babylonian numbers were positional. For our system, '1' can mean one; in the next position, 1 times the base ten, as in 10, or one times the base 10^2 and so on. However, the Babylonian base was sixty (the origin of our seconds and minutes). Larger numbers depended on their positions from right to left, so that a ⊤ in second position meant 1×60, ⟨ in the second position meant 10×60, ⊤ in third position meant 1×60^2 (3600) and so on.

This system of writing numbers had clear practical purposes. Babylonian accountants wrote down numbers to count stock, pay workers and manage trade. Babylonians also needed to calculate and record calculations for building and for astronomy.

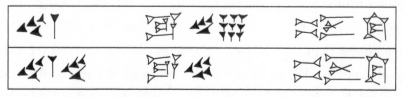

$(4 \times 10) \times 60 + 1 = 2401\ (49^2)$ $(4 \times 10 + 1) \times 60 + 40 = 2500\ (50^2)$

Figure 1. 1A. Cuneiform symbols for one and ten. This is a positional notation so that the symbol for one can also mean the Babylonian base sixty. 1B. These are examples from a table of squares and square roots discovered by Sir Henry Rawlinson in 1855. It reads from left base (60) to right base (1) showing cuneiform the equivalent of 49^2 and 50^2.[1]

Even earlier, the inhabitants of the Fertile Crescent – parts of modern Iraq – began to practise settled agriculture around 12,000 years ago. This prompted the need to record the quantities of different types of produce, year by year. Initially, the local Sumerian farmers, or their accountants, invented a system of clay tokens to indicate the quantity of different products. Extensive archaeological deposits of these tokens have been found in the Mesopotamian city of Uruk in what is now Iraq, and Susa in modern Iran, dating from before the invention of writing around 5000 years ago. Examples and their numerical values are given in Figure 2A. Some of the produce was traded quite widely, but the producers could not always travel with their goods, so they needed a system to ensure that the goods as sent were received by the recipient, who could then pay for them. For this was needed a bill of lading, a document issued by the carrier to acknowledge receipt of the cargo for shipment, and an invoice to tell the recipient what was sent, and therefore the cost to be paid. To do this, the tokens, Figure 2, were first enclosed in a clay envelope, called a *bulla*. Of course, the transportation company could, if nefarious, open the envelope, remove a token or two, and remove the corresponding amount of goods for sale on their own behalf, and the recipient would then only pay for the remaining goods and the sender would be deprived of the rightful payment. To counter this problem, the clay envelopes would be marked with symbols to denote the tokens contained within. It was then realized that the tokens were not really necessary and the symbols alone impressed into a clay docket would constitute the invoice. In fact, complex transactions could be impressed into a clay tablet indicating the objects transacted and to whom, along with information about quantities.

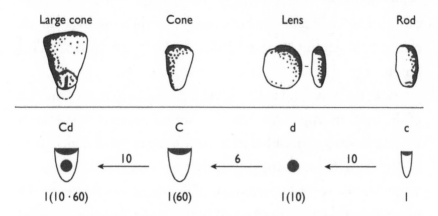

Figure 2. The earliest written numbers from Sumer, perhaps ten thousand years ago. Tokens used in proto-cuneiform/proto-Elamite transactions with their numerical values: 1, 10, 60. The 10 symbol within the 60 symbol in the left-hand token indicated 10 × 60. There were also symbols for the powers of 60.[2]

A great deal of book-keeping information can be embodied on a small clay tablet. One proto-cuneiform tablet recorded the distribution of barley to four officials on one side and the titles of the relevant officials on the other.[3]

The reason the Sumerian accountants used base sixty is mysterious. Theon of Alexandria (fourth century BCE) suggested that it was the lowest of all the numbers that had the greatest number of divisors – 2, 3, 10, 12. It may have had something to do with a year having (approximately) 360 days, and hence 360°. The historian of mathematics Georges Ifrah relates it to the number of phalanxes on a hand, twelve, thus each finger represented a multiple of twelve: 12, 24, 36, 48 and 60.[4] Base sixty turned out to be very useful for us.

Many centuries later in another part of the world, across thousands of miles of ocean, two other counting systems were being developed independently, Mayan and Inca. The Maya carved their elaborate numerals on stone, while the Inca empire recorded important numerical data in knots on strings, called quipu.

Both the Maya and the Inca had big empires to run: in the case of the Inca, an empire that stretched from Colombia in the north to Chile in the south, with the king (the Inca) in the capital, Cuzco, in Peru. This meant raising taxes and armies, and that in turn meant supplying the central government with annual inventories of the region's crop yields and taxes, and censuses of population by social class, births, marriages, deaths, and of men fit for combat. In the words of one contemporary, Garcilaso de la Vega (1539–1616), the son of a Spanish conquistador and an Inca princess:

> The Inca [king] had a census made, not only of the inhab-
> itants of each province, but also of all that these provinces
> produce annually in the way of goods of every sort. This
> was in order to learn what provisions would be required to
> come to the assistance of his vassals, if they were to suffer
> from shortages, or had a poor harvest, and the quantities
> of wool and cotton that would be needed to clothe them.[5]

The central government assigned two *quipucamayocs* ('keepers of knots'), specialists who could read and write quipus (some-times written as 'khipus'). One was responsible for sending the revenue accounts and the other for records of the residents. Although the Inca in Cuzco spoke Quechua, the empire con-tained speakers of many languages, and quipus were independent of language, like our Arabic numerals.

How the meaning of quipus was discovered is itself a fascinating

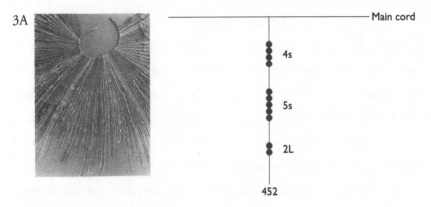

3A

Main cord

4s

5s

2L

452

Figure 3. Inca quipus. 3A is an example from the collection of the National Museum for Anthropology and Archaeology in Lima, Peru. It is a record of market transactions. The basic principle is a base-ten system with highest power of ten nearest the main cord. 3B is from an analysis of the quipu code showing that it is a vertically oriented positional decimal notation.[6]

story, and its code was presumed as lost as Linear B, which took over fifty years to decipher and the work of many minds until an amateur, Michael Ventris (1922–1956), finally cracked the code. Cracking the quipu code took until the 1920s, when the American scholar Leland Locke (1875–1943) decoded the numerical content of the quipus. The difficulty modern scholars have had in deciphering them suggests that the code used in quipus was actually quite complicated and difficult to learn.

The coding in quipus can be highly complex. For example, strings above the main cord can represent the totals and the context. The lower cords can record not only numbers, but actual item prices, the tax paid, and even the change.[6]

It turns out that the coding is even more sophisticated and subtle than it first appeared to Locke. Sabine Hyland of St Andrews University, scholar of all things Peruvian, discovered that the way the cords were twisted encoded the group making it.

Thus S ply (clockwise) corresponded to one group and Z ply (anticlockwise) corresponded to another group.[7] Using an old text and the corresponding quipus, Hyland discovered quipus that recorded milking cows such that 'The 60 cows on the first cord with a final Z ply are dry . . . [those that are not milked], and the 170 cows on the other cord with a final Z ply are not milked daily . . . the 85 cows on the cord with the final S ply were milked daily. On these cords of the khipu, the direction of ply corresponds to the milking status of the cows rather than gender: final S ply = milked.[8] In fact, Spanish witnesses of the Inca period said that quipus encoded narratives and were sent as letters, and indeed contemporary indigenous Andeans claim that their quipus are sacred epistles describing warfare.

Like Roman and Greek alphabetic numerals, quipus didn't lend themselves to easy calculation. Hyland found that

> When economic and/or tribute data was read off of khipus, calculations were done on the ground using . . . pebbles or grains of maize arranged onto a grid known as yupana, even into the twentieth century. There are many colonial-era references to this. In a Peruvian archive I found an unpublished transcription that an anthropologist had made of a conversation he had with a khipu expert in 1935 – the expert mentioned that they still did calculations with kernels on the ground when they used khipus to settle the accounts. (personal communication; see Figure 4)

Nevertheless, according to Garcilaso de la Vega, 'The Incas also had an excellent knowledge of arithmetic and the way they counted was quite remarkable . . . [they calculated] with little pebbles and

Figure 4. A *quipucamayoc* with a quipu and a counting board ('yupana'), from *Nueva corónica y buen gobierno* (*The First New Chronicle and Good Government*), written by a Quechua nobleman, Felipe Guaman de Ayala (*c.*1535–after 1616).

grains of corn, in such a way that there could be no mistakes in their calculations.'[5]

The Inca administrators seem to have adapted an earlier system. Hyland pointed me to an ancestor of Inca quipus:

The Wari Empire, which lasted from about 600 AD to about 1100 AD, is the first Andean civilization known for certain to have possessed khipus, although their structure is different from that of Inka khipus. Most of the Wari khipus appear to have what we call a 'loop and branch' structure . . .

One of the primary characteristics of the Wari khipus is that the pendant cords are wrapped with sequences of colour . . . [but] most Wari khipus do not seem to have encoded numbers, although there may be some instances where they did. (personal communication)

Nevertheless, there seemed to be training in quipu reading and writing, though it is not clear how widespread these skills were. Hyland told me that:

The Incas had schools in Cuzco where they taught how to read and write quipus. Sons of local leaders were required to attend these schools, so they would spread what they had learned back to their home communities. The Inca government did send quipu experts throughout the empire, although most scholars think this was to collect the information that the government needed about the population, tribute, etc. But they could have taught reading and 'writing' in quipus as well.

Jeffrey Quilter of Harvard University pointed to a very recently discovered and potentially very exciting clue to an earlier base-ten number system in a lost language in coastal Peru. Under a collapsed church at Magdalena de Cao Viejo, Quilter and his colleagues found a letter in Spanish on the reverse of which 'traces of a lost language were written'. The writer had noted 'the Spanish names for the numbers 1–3 in a single column, then changed to writing Arabic numerals for the rest of the sequence on the left-hand side of the page.' Each number was accompanied by a word which we interpret as the equivalent number in a native language. After the numeral ten, the sequence continues with 21, 30, 100, and 200. Both the format and the contents of the brief list suggest that the author may have been recording numbers with the aim of understanding the numerical system in question, possibly during or shortly after an interview with a native informant.[9]

So here we have a clue to a ten-based counting system that may have been widespread in the Andes before the Incas, before anyone, thought to record it.

Even earlier than the Incas, the Maya of Central America were using a written counting system that included a symbol for zero, several hundred years before the Indus Valley civilization invented their symbol, and exported it to Europe in the thirteenth century. The Mayan system was also positional, but used base five represented by a horizontal line. Thus nineteen is represented as ≡, four dots on top of three fives. Beyond nineteen, it used base twenty in a vertical positional system of 1s, 20s, 20^2 and so on (see Figure 5, overleaf).

In fact, the Maya had two systems of recording numbers. The one shown in Figure 5 was to keep track of commercial dealings, and merchants used counters or cacao beans, laying them out in their proper positions to make their calculations. The second was for calendrical calculations based on 360 days. 20 days = 1 uinal; 18 uinals = 1 tun, 360 days; 20 tuns = 1 katun, 7200 days; up to 20 kinchiltuns = 1 alautun (23,040,000,000 days).

The Maya calendar was far more complex than ours, for it served a variety of purposes, both practical (such as

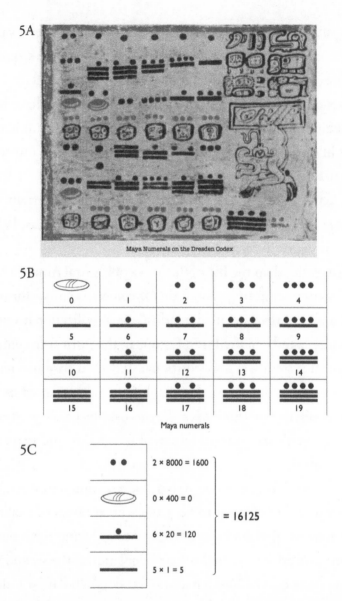

5A

Maya Numerals on the Dresden Codex

5B

0	1	2	3	4
5	6	7	8	9
10	11	12	13	14
15	16	17	18	19

Maya numerals

5C

2 × 8000 = 1600

0 × 400 = 0

6 × 20 = 120   = 16125

5 × 1 = 5

Figure 5. Mayan counting and calculation. 5A is a page from the Dresden Codex, the oldest surviving document from the Americas, dating to the thirteenth or fourteenth century, rediscovered in the city of Dresden in Germany, hence the book's present name. It shows many examples of counting and numerals. 5B shows that Mayan numerals use a base of twenty (fingers *and* toes), with a shell indicating zero. It is also positional, like ours, but organized vertically instead of horizontally.[10]

determining times to plant maize) and esoteric (such as plumbing the mysteries of astrological divination) . . . The Maya kept records of recurring time based on the movements of celestial deities (sun, moon, and the planet Venus, being the most prominent) . . . The three cyclic counts most frequently used by the ancient Maya – the 260-day sacred almanac, the 365-day vague year, and the 52-year calendar round – are very old concepts, shared by all Mesoamerican peoples.[11]

There is a curious similarity between fourteenth-century Inca quipus and the much older Maya system. First, they both use bases, the quipus our familiar ten-base system, and the Maya a twenty-base system. Second, both systems are organized positionally. This is not surprising, since as we have seen, the Sumerians' sixty-base system was also organized positionally, but it is important to realize that base systems do not have to be positional. Our ten-based number name system is not positional. It gives distinct names to the powers of 10 – *ten, hundred, thousand* – as does the Chinese system – 10^1 *shí*, 10^2 *bǎi*, 10^3 *qiān*, 10^4 *wàn* . . . Third, and perhaps most intriguingly, positions are organized vertically, whereas the Hindu-Arabic ancestors of our own system are organized horizontally, in ours left to right. In the Maya system, the bases increase downwards, and in quipus the highest base is nearest the main cord, that is, the bases increase upwards. Could the older Maya system have influenced the early inventors of numerical quipus? I could find no evidence of this, but there was trade between the Inca realms and Central America through Costa Rica both by sea from the Atlantic side and by land on the Pacific side. However,

Jeffrey Quilter, discoverer of the 'lost language' of Peru, as well as an expert on pre-Inca Peru, is not convinced by this speculation:

> The main issue about contact between the Maya and Inca is that Classic Maya culture, when writing and numbering systems were most elaborate, was from about AD 250 to 700, whereas the Inca as a political power were active from about 1300 to 1500. There were Postclassic Maya in Mexico and Central America when the Inca were in power, and they did write, but the heyday of Maya writing as seen on the temple and stelae carvings you probably are thinking about are all in the Classic Period. (personal communication)

Similarly, Hyland sees quipus deriving from other braiding technologies and uses: 'I suspect that khipus may have derived from slings – Andean slings are considered the most complex in the world in terms of their construction and braiding' (personal communication).

These notations – Sumerian tokens, cuneiform, Egyptian and Maya characters, and quipus – are all symbolic. An obvious question is why these cultures should have invented symbols if they did not already possess the concepts they wanted to symbolize. We know that they already had words linked to these symbols. What about earlier attempts to notate the results of counting, where, perhaps, no counting words existed? This suggests that these cultures were seeking to notate concepts they already possessed.

Prehistory

In the Stone Age, before about fifteen thousand years ago, when there was no writing, did anatomically modern humans, or their ancestors, try to notate the results of counting? Did they create what the archaeologist Francesco d'Errico terms an 'artificial memory system' to record the results of their counting?[12] If they did, this would be an indication that they already possessed the idea of counting and of the number of objects in a set. Maybe followers of the first teacher of recording counts in an artificial memory system were simply following a learned procedure without understanding the idea, rather like my following a recipe in a cookery book but without understanding why I have to leave the meat to stand for ten minutes before carving and serving.

Bones and stones: one-to-one correspondence

A fundamental idea for all counting is to enumerate each object to be counted once and only once. To create a record of the count, the simplest way is to make one mark for each object. This underpins the understanding of sets and their numerosity, and it also underlies the acquisition of verbal counting. Matching two sets in one-to-one correspondence ensures that both sets have the same numerosity (see the Gelman and Gallistel 'counting principles' in Chapter 2). This is the idea behind tallies, and it turns out to be almost universal in the development of record-keeping.

Even when humans invented symbolic notations, the first few numerals are simply tallies. As we saw, the ancient Sumerians made one mark for each object recorded up to ten, and similarly with cuneiform, the one symbol is simply repeated to ten. Even Chinese and Japanese, with their sophisticated counting word systems, write numerosities up to three by three marks. Our clocks notate hours to three with one to three marks. Within living memory, tallies have been used by Pacific Islanders and Swiss herdsmen.[13] Tallies can be complex and positional. For example, on the counting boards widely used in Roman times and still used in schools, you tally ones, tens and hundreds in separate columns.

Even the Bank of England used tally sticks until the practice was abolished by statute in 1783. However, they were so useful that they continued to be used until at least 1825. According to the historian of money C.R. Josset, cited by Graham Flegg,[13]

> Some of the tallies represented payments to government departments . . . one of these . . . is eight feet six inches long, and represents the largest amount to which one tally was restricted – £50,000. It owes its existence to the fact that it had never been returned to the Exchequer, as the payment represented a government debt which had never been repaid.

Recent history reveals the use of *double tally sticks* which consisted of one piece of wood cut in two, with identical tallies on each piece, with the borrower keeping one and the lender keeping the other. This makes cheating impossible. They were in use in Central Europe until quite recently.[13]

The word 'tally' itself comes from the Latin *talea*, which means cutting and stick, and is the source of the French *taille* meaning cut or size and the Italian *tagliare*, also meaning to cut. The Latin word *putare* means to cut, as in *amputare* and even *computare*, which means to count; and our word 'score' also means to cut and to keep count. These etymologies imply the shared historical roots of making marks, tallying and counting.

The simplicity of tallying, and its foundational role in counting and arithmetic, has a much deeper history. Our Ice Age ancestors more than twenty thousand years ago made marks on bones and stones that have survived the centuries. They may have also notched sticks as well, but these haven't survived. But what were they counting and why? They weren't agriculturalists with a surplus to trade, like the Sumerians. From what we know of modern hunter-gatherers, trading was typically carried out face to face, so there was no need of invoices or bills of lading.

This doesn't mean that prehistoric tallying is simple and static. Like other human skills, it developed in a sequence of stages, becoming increasingly subtle and sophisticated.

Here I follow the work of Francesco d'Errico from the University of Bordeaux, France, and the Centre for Early Sapiens Behaviour in Bergen, Norway, who gave a brilliant and inspiring talk at a meeting on the origins of numerical abilities of the Royal Society in London in 2017. D'Errico is one of the key figures in archaeological research on the mind of ancient sapiens even though he is, in a sense, an accidental cognitive scientist. His original studies in Turin were in anatomy, particularly in using microscopy to study bone modifications, but when he moved to the Institute of Human Paleontology in Paris, he was asked to help with an exhibition at the Musée de l'Homme on Upper Palaeolithic art. While he was looking at notched and engraved objects in the displays he was preparing, in particular at the abstract patterns bearing pebbles from the Azilian era (about ten thousand years ago), he realized that he could apply the expertise gained in Turin to better understand how these pebbles have been engraved and for what purpose (personal communication). He subsequently used his skill in microscopy to examine claims made on the basis of marks on bones and stones. I'll mention one particular example below.

I outline these stages in Table 1. You will see that the development from one stage to the next is slow, much slower than technological developments in more recent times, and spectacularly fast in the past few years. So why were Stone Age humans slow to improve their technology? There is a simple answer to that: population density, as I will try to explain. Fascinating research by archaeologists and geneticists at UCL, Adam Powell, Stephen Shennan and Mark Thomas, asked why the development of technology of the Upper Palaeolithic, the Late Stone Age, occurred so much later – forty-five thousand years later – in Europe than in Southern Africa. Had there been a change in the brains and the cognitive capacities in Southern Africa that gave them an advantage and these slowly spread into Europe and Asia, bringing with them what is sometimes called the 'modern package'? The package includes:

> symbolic behavior, such as abstract and realistic art and body decoration (e.g. threaded shell beads, teeth, ivory, ostrich egg shells, ochre, and tattoo kits); systematically produced microlithic stone tools (especially blades and burins); functional and ritual bone, antler, and ivory artifacts; grinding and pounding stone tools; improved hunting and trapping technology (e.g. spear throwers, bows, boomerangs, and nets); an increase in the long-distance transfer of

raw materials; and musical instruments, in the form of bone pipes.[14]

The other question is why there was a delay of a hundred thousand years between the emergence of anatomically modern humans that evolved in Africa between 160 and 200 thousand years ago and behaviourally modern humans. The critical factor at this period, and I am guessing today too, is population density, in particular the density of communication. Think of it like this. Mr or Ms AMH (anatomically modern human) invents a new technology, for example microliths – stone blades – that are good for cutting food. For this invention to survive and become part of the package, someone else has to learn how make these stone blades. If Mr or Ms AMH is a member of a small group, there may be no one smart enough, or keen enough, to learn how to make the blades well. In a larger group, there will be a better chance of finding a suitably competent apprentice. That is, the probability of high-fidelity transmission of the new technology will be greater if the inventor is in a larger group. Powell and his colleagues can trace the emergence of elements of the package to the population densities in Africa, and when AMH moved into Europe and Asia in those locales too. The population density in Europe increased rapidly, as did the technologies, but Southern Africa had a head start. And these new technologies could well have had a feedback effect, so that the groups who had more of the package may have increased their populations faster than those who had less.

Humans were thus making marks as a by-product of some other activity, such as cutting meat from a bone; that is, they were not tallying. According to d'Errico's analysis, they co-opted this

3 million years ago	Unintentional cut marks	
900– 540 thousand years ago (kya)	Coherent abstract patterns	Engraved freshwater mussel from Trinil, Java (*Homo erectus*)
100–45 kya	Sequences of individual marks	Les Pradelles, France, incised hyena femur, 72–60 kya
44 kya	Marks added over time and grouped	Notched baboon fibula, Border Cave, South Africa, 44 kya
40 kya	Complex codes	Incisions on the Blanchard, France, ivory spatula 36kya

Table 1. Five stages in the development of Stone Age counting.[15]

activity in order to record numbers – to tally, in other words. It doesn't stop there. They also began to group their markings. The originality of d'Errico's approach is to rely on microscopy to establish which marks were made by which tools and in which way, for example the notches made forty-four thousand years ago by each tool on baboon fibula.[15]

By using microscopy it is also possible to tell whether marks were made by the same tool or by different tools. You can test this today, by asking skilled or unskilled humans to make marks on bones or stones.

When d'Errico examined the incised pebbles, he found that:

all the incisions on a given pebble were produced by the same tool in a single series of operations. This finding suggests that these engravings were not, as has been supposed, lunar calendars or hunting tallies, since these would have been created in several distinct operations over a relatively long period and probably with more than one tool.[16]

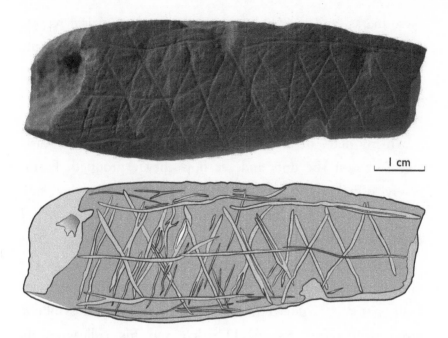

1 cm

Figure 6. From the Blombos Cave, S Africa, 73 kya. *Homo sapiens* were making coherent abstract patterns on pieces of ochre.[17]

Cave markings

Marked bones and stones are able to survive thirty or more thousand years of exposure, but ancient humans also marked cave walls. The depictions of animals on the walls of European caves such as Lascaux, Altamira and Chauvet are deservedly famous for the painters' skill and artistry. But there are other marks that are not depictions of animals or humans that are more difficult to interpret.

Twenty years ago, when I saw a row of red dots on the wall of El Castillo cave in northern Spain, of course I thought these could be tallies of something. The information then available was no help. I was not the first, of course, to notice these marks, and they have appeared in archaeological reports for more than a hundred years, and dots themselves, on the basis of the analysis of the pigments, have been described as the earliest European cave paintings made forty thousand years ago.

In my diary of the trip to El Castillo, I wrote, I'd dearly love to count the dots and see how they're grouped. Fortunately, d'Errico and his colleagues have done just that, and of course with their technical expertise much better than I could have done.[18] Differences in pigment texture and composition suggest that the dots were not necessarily all made by the same person, for the same reasons, or even using the same symbolic system, if indeed their makers were using a symbolic artificial memory system. D'Errico and his colleagues do not speculate on what these dots mean, but I am happy to believe, until proved wrong, that the dots represented counting something.

On the wall of the Chauvet Cave in in the Ardèche Gorge in France, there are mysterious marks in a deep red colour that comes from heating the ochre, which is otherwise yellow. This suggests that the marks were made deliberately and for some purpose. One group consists of three rows of three marks: two groups of ochre dots and one group of ochre lines. Was this a coincidence or did the maker intend the same number of marks in each row?

D'Errico says that it is important to see dots like this in relation to depictions of, say, animals. Could they represent the number of animals killed, eaten or just observed in a particular place?

Figure 7. Figures from Niaux Cave, drawn about fifteen thousand years ago. This image has been interpreted as a dead bison with, to its right, a tally of the number of hunters who killed the bison – the circle of dots around a single central dot. The lines with bumps, called 'claviforms', have been interpreted as women. The dots and line in the centre remain even more mysterious.[20]

Figure 7 is a more recent example (maybe fourteen thousand years ago) from the Magdalenian period in Niaux Cave in France, contemporary with the cave drawings in the famous caves of Lascaux and Altamira.

Here we see fourteen dots in a circle surrounding a single dot. Does the number of dots represent the actual number of bison encountered? This would be easier than drawing fourteen bison. There are many examples of ancient rock art where a number of humans or animals are depicted. Does the group depicted represent exactly the number of people present? That is, are these one-to-one tallies of the objects depicted? Similarly, groups of handprints are found on cave walls all over the world; is there one hand for each person present?

Neanderthals

What about other members of the genus *homo* – for example, the Neanderthals, *Homo neanderthalensis*? They appear to have descended from an earlier human species, *Homo erectus*, between five hundred and three hundred thousand years ago. They had brains at least as large as *Homo sapiens*. They have long been treated as our dim cousins: no art, no language, no religion, poor tools. Ernst Haeckel (1834–1919), an immensely gifted and influential zoologist and naturalist, as well as a 'scientific racist', called these stocky creatures, with their heavy eyebrow ridges and sloping foreheads, *Homo stupidus*. It was also widely assumed that when anatomically modern humans entered Europe they replaced them because they were smarter, had better tools, had language and were capable of conceptual thinking, as shown in their portable art and cave art.

We've recently had to rewrite the history of Neanderthals. First, Svante Pääbo and his colleagues at Max Planck Institute of Evolutionary Anthropology discovered that all humans – whose ancestors left Africa and migrated to Europe, through Asia and down through Indonesia to Australia – have traces of Neanderthal DNA. Maybe I myself have inherited a Neanderthal genetic variant that explains my Type 2 diabetes. Pääbo reports that he gets a lot of emails from men who say that they are Neanderthals, but almost no women say that they are. However, a lot of the women tell him that they are married to a Neanderthal.[21]

Second, when evidence of elements of the 'modern package' – such as beads and shells for body decoration, bone, antler, and ivory artefacts, and art – was found in Neanderthal sites, it was

assumed that they learned the skills or concepts from neighbouring *Homo sapiens*. However, it has recently been discovered that
Homo neanderthalensis didn't just learn to create the modern
package from the supposedly more advanced *Homo sapiens*. We
know this because of exciting new methods of dating. Radiocar-
bon dating, which depends on organic materials, such as charcoal
or bone, is unreliable beyond about forty thousand years. The
new method uses the ratio of uranium to thorium found in calcite (calcium carbonate, $CaCO_3$). The uranium decays into
thorium over time, so the higher the proportion of thorium, the
older the substance. Alastair Pike of Southampton University
used this technique to show that calcite crusts that formed over
ochre paintings in La Pasiega cave in Spain were sixty-four thousand years old, meaning the paintings were even older, and at
least twenty thousand years before anatomically modern humans
appeared in Europe.[22] That is, these works had to have been created by Neanderthals (see Figure 8, overleaf). Since then, several
other sites have yielded evidence of their work older than sixty-
four thousand years. These paintings are the oldest dated cave
paintings in the world.

In support of the idea that Neanderthals counted is a study of
a fragment of hyena femur from Les Pradelles cave in France by
the redoubtable Francesco d'Errico and colleagues. The incisions
on the bones look, to my untrained eyes, very much like those in
Figure 6 (p. 103). Moreover, on a raven bone Neanderthal craftsmen
or craftswomen made incisions 'with the intention of producing
equidistant notches', with the same accuracy as modern humans
tasked with doing just that.[12]

There is another extraordinary piece of evidence that Neanderthals counted. In 2020 the paleoanthropologist Bruce Hardy

Figure 8. A rendering of the design showing the pattern of dots clearly the dots and depictions of an auroch head and rear. There is also an intriguing and mysterious design on the right.[22]

of Kenyon College in Ohio and an international team of experts discovered that Neanderthals in Abri du Maras (France) made ropes around fifty thousand years ago.[23] Why is this so extraordinary? The fibres come from the bark of trees, and this kind of organic matter almost never survives in the habitations of our ancestors from this long ago.

These cords were highly complex constructions, with the component yarns made with clockwise 'S-twist', and then three strands of yarn were bound together in a counter-clockwise 'Z-twist', creating a highly durable cord that won't easily unravel.

The mere fact that Neanderthals had twisted fibre technology enormously extends the scope of their industry. They could use these cords to make clothing, rope, bags, nets, mats and even boats. The production of the cord from Abri du Maras requires keeping track of multiple, sequential operations simultaneously. This is not just an iterative sequence of steps because each has to have access to the previous stages. Thus, Hardy and colleagues conclude,

> Understanding and use of twisted fibres implies the use of complex multi-component technology as well as a mathematical understanding of pairs, sets, and numbers . . . Cordage production entails context sensitive operational memory to keep track of each operation. As the structure becomes more complex (multiple cords twisted to form a rope, ropes interlaced to form knots) . . . requires a cognitive complexity similar to that required by human language . . . It is difficult to see how we can regard Neanderthals as anything other than the cognitive equals of modern humans.

Homo erectus

Long before Neanderthals emerged in Europe, or anatomically modern humans left Africa, ultimately to settle in Australia and the Americas, *Homo erectus* also left Africa and reached Dmanisi in Georgia (1.77 million years ago), Java (1.5 million years ago), northern China (700,000 years ago) and probably Europe (900,000 years ago). And the adventurous *Homo erectus* people who reached Trinil in Java 430–540 thousand years ago incised the shell in Figure 9. It looks a bit like the Blombos stone (Figure 6, p. 103), and could be the coherent abstract pattern that is an artificial memory system, but of course we can't tell if its creation involved counting.

The marks on the shell in Figure 9 look very much like the marks on the definitively human marks in Figure 6, and they are the oldest geometric engravings found so far. We know these marks were intentional because this example comes from a collection of shells from the same site with holes in them that could only be made by hominins, but at present we have no idea what they mean. Nevertheless, 'The manufacture of geometric engravings is generally interpreted as indicative of modern cognition and behaviour,' and so we must ask whether this is confined to *Homo sapiens* and *Homo neanderthalensis* or whether our more

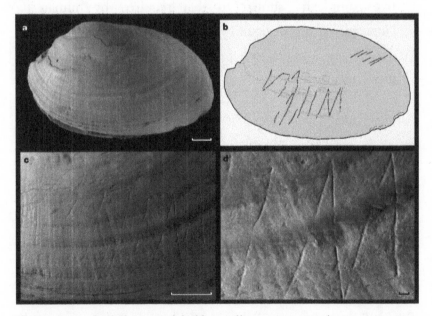

Figure 9. A shell decorated half a million years ago by our ancestor *Homo erectus* found on the island of Java in Indonesia.[24]

distant ancestors also possessed these cognitive capacities.[24] Very similar markings have been found on a mammoth bone in Bilzingleben, Germany, also from about half a million years ago, supporting the idea that *Homo erectus* was making some kind of record that may have involved counting.[15]

These marks may be a very early artificial memory system, and each mark may be the count of one object observed. Alternatively, they may just have looked nice to the maker.

The markings on bones, stones, shells and cave walls suggest that our ancestors, members of the genus *homo*, created artificial memory systems that could have been used to record the results of counting. One plausible explanation is that these markings were tallies in one-to-one correspondence to things counted. They could be bison seen or killed, participants present at a ritual, or phases of the moon. Perhaps things mysterious to us were counted, too. In the deepest parts of caves in Stone Age Europe, strange chimeric figures decorate the walls, such as the 'sorcerer' in Les Trois Frères cave in France, which is part man, part stag, part horse with owl-like eyes.[25]

Stone Age counting words

Did our pre-literate ancestors use counting words? Of course, we have no recordings of the earliest speech to demonstrate these changes, but it is now possible to reconstruct ancient languages if we have documentation of different branches of a language family and a time-stamp for forms of the word for a given meaning. We can now determine the oldest words in continuous use in unrelated language families.

Most of us have heard about Indo-European, the source of

Latin, Ancient Greek, Sanskrit and almost all European languages and many Indian languages. These were all base-ten languages and their counting words are very similar. These similarities have led many scholars to reconstruct a common original language, Proto-Indo-European (PIE), hypothesized to have been spoken in one place in the Late Stone Age (Neolithic) from which they all derived and evolved in a rule-governed way, as the speakers spread out over Europe and Asia, rather as Darwin's finches evolved slightly different forms as they settled on different islands in the Galapagos.

Take the word for two. French: *deux*; German: *zwei*; Welsh: *dau*; Latin: *duo*; Greek: *duo*; Russian: *dva*. More distantly, Sanskrit: *dva*. All these forms are derived from the PIE, something like *du(w)o*.

For three: French: *trois*; German: *drei*; Welsh: *tri*; Latin: *tres*; Greek: *treis*; Russian: *tri*. More distantly, Sanskrit: *trayas*. PIE something like *treyes*.

These words are cognates of each other and of the original PIE. When we get to a hundred, things get more complicated, but still, it is argued, cognates. There is one line of which Sanskrit and Slavic languages, such as Russian, are examples. These are called the 'Satem' languages where the first sound for 100 is *s*, as in the Sanskrit *satam* and Russian *sto*, derived through several well-established sound changes from the proposed PIE *k̂mtóm*. The first sound *k̂*, a palatal stop, becomes the palatal fricative, *s*. For ten, the Sanskrit is *dáśa* and the Russian *desatj*, derived from the proposed PIE *dék̂m̥t*. The other descendants of PIE are the 'Kentum' languages, Latin, Greek, Celtic and the Germanic languages among others, because the Latin for 100 is *centum* pronounced *kentum*, and 10 is *decem* pronounced *dekem*, evolving from the *k* in *ḱmtóm* and *déḱm̥t*. The *h* in hundred comes through additional sound changes.

These reconstructions of the common ancestor of modern European and Indian languages, and Tocharian, Armenian, Albanian and several extinct languages of the Near East (though not Basque, Finnish or Hungarian), suggest that number words are both early in human language history and stable in not changing much and evolving in predictable ways. Thus, humans – the speakers of PIE – were counting with counting words at least nine thousand years ago, the most recent date for PIE.

These words have evolved slowly over time according to well-understood principles of sound change. There are now ways to reconstruct other language families that are even deeper in human history to see whether number words can be traced back even further. Can we find counting words in unrelated language families such as the Bantu languages of Africa, which is a group of around five hundred distinct languages, including Zulu, Xhosa and Swahili, with many millions of speakers? Or in the Austronesian languages of the Pacific, spoken by 400 million people spread out over enormous distances from Madagascar to Polynesia?

In these language families, do number words have equally ancient roots? That is, are modern words cognates? And are they as ancient as words for concepts that would certainly have been familiar to the earliest speakers, for example, *mother*, *rock*, *fire*, *child* or *hand*? Take *hand* in the descendants of PIE. It turns out that there isn't just one form changing slowly by the principles of sound change. In fact, quite the opposite. There are two quite different forms, one derived from something like *gʰésr̥* with its descendants in *hand* in English and German, *hás-ta* in Sanskrit.

However, another, quite different and unrelated form emerged in the Romance languages, *manus* in Latin descendants, *mano* in Spanish and Italian. Somewhere along the line descendants of $g^h\acute{e}sr$ were replaced by the *man-* form.

Now, it's possible to look at the replacement rates of different words. One wouldn't look for words meaning *computer* in the Stone Age, but for more basic ideas. The linguist Morris Swadesh (1909–1967) created a famous list of about two hundred basic vocabulary items – actually basic meanings – that were independent of particularities of culture, climate and setting, so that he could compare languages and plot their changes. It included words for people (woman, man, child), kinship terms (wife, mother), the names of body parts (hand, tongue), actions, pronouns, adjectives (good, bad, dirty) and the number words from one to five.

Studies by Mark Pagel and Andrew Meade and their colleagues compared replacement rates of items in this vocabulary. They found that

> *dirty* was the most rapidly evolving word in the [Swadesh] list, with a rate of lexical replacement of about 0.0009 per annum, or approximately one new non-cognate form every thousand years. This rate of replacement yielded 47 different non-cognate forms among the 86 Indo-European (IE) languages in our sample. By comparison to words for *dirty*, the words with the slowest rates of lexical replacement were represented by just a single cognate form across the entire IE language tree. Among these slowly evolving forms were the number words *two*, *three*, *five*, and the pronouns *who* and *I*.[26]

They then turned their attention to the Bantu language family (103 languages) and the Austronesian language family (400 languages). They used historical records for each item to establish (where possible) dates of each item, and then applied very fancy statistics to reconstruct 'time-calibrated' trees. 'The IE tree is dated to approximately 7654±915 years old, the Austronesian tree to 6924±500 years and the Bantu tree to 6929±418 years.' It turned out that each language family had very similar rates of lexical replacement.

They then plotted the words with the slowest rates of change – that is, where a word is replaced by a non-cognate (see Table 1).

Although these reconstructions are based on about ten

rank	Indo-European (n = 200 words)	Bantu (n = 102 words)	Austronesian (n = 154 words
1	two	eat	child
2	three	tooth	two
3	five	three	to pound/beat
4	who	eye	three
5	four	five	to die
6	I	hunger	eye
7	one	elephant	four
8	we	four	ten
9	when	person	five
10	tongue	child	tongue
11	name	two	eight

Table 1. Rank order of rate of lexical replacement for the eleven meanings with the slowest rates of change; rank = 1 is slowest. Words 'one' to 'five' in italics. The probability of all five low-limit number words appearing in the slowest eleven for IE is 0.0000002; the probability of four of the five low-limit number words appearing in the slowest eleven is 0.00036 for Bantu and 0.00007 for Austronesian.[26]

thousand years, Pagel and colleagues argue that these low-limit number words – up to 'five' – can remain stable over a hundred thousand years because of the slow rate of change. Numbers above five are not preserved in the same way, suggesting that they may have been added later.

Unless there was some kind of Ursprache that was the first language from which all these language families descended, these findings suggest that there wasn't a single inventor of counting words to support his or her counting practices; rather, these words were used to express concepts that their speakers already possessed at least ten thousand years ago, and probably a hundred thousand years ago. There would be no point in having a word for 'five' unless the speakers already had the idea of fiveness.

Of course, this doesn't rule out the possibility that there was a first teacher who invented counting, or several very first teachers independently inventing counting and the corresponding count- ing words and teaching them to their small, but possibly expanding, clans, but it seems unlikely.

As I noted in Chapter 2, not all languages have counting words. They are missing in some Amazonian languages and most indigenous Australian (Pama-Nyungan) languages. In fact, most of these languages appear to have number words just for one, two and three, sometimes four, and in nine languages only words for one and two.[27] Even these words, in the languages I know about, and perhaps most, are not used for counting. For example, in Warlpiri, historically, and like many other Australian languages, the counting system consisted of 'none', 'one', 'two', 'few/several' and 'many.'[28] Even 'one' and 'two' seem to be used rather loosely, so that 'two', for example, may denote 'three' as well. These num- bers are often also expressed as grammatical markers on nouns and other parts of speech, depending on the language. In Warlpiri, there are markers for singular, dual (= two), paucal (few) and plural (many). In Warlpiri, in fact, like forty-nine languages in an extensive corpus compiled by Claire Bowern and Jason Zentz, three is represented compositionally as 'two and one'. More recently, and with the help of linguists such as Mary Laughren, working to improve education, Warlpiri school kids are being taught new number words.

The situation is very similar with remote Amazonian languages. Alexandra Aikhenvald (see p. 49) writes:

> The lack of counting as a practice does not mean that people cannot perceive differences in quantities. Holmberg remarks that the Sirionó of eastern Bolivia 'are unable to count beyond three'. He goes on to say that they are perfectly able to notice if one ear of corn has been removed from a bunch containing a hundred ears . . . [But] 'since the principle which underlies counting is present, filling the gap is a rather trivial matter' (Hale 1975). And this is why Spanish or Portuguese counting systems are quickly mastered and used by the Amazonian peoples, especially in situations where the use of money is important.

It is not clear why these languages should lack counting words. Here we have to ask what ancient people counted, and why. The development of Sumerian tokens and incised marks on clay was needed by Sumerian accountants to log the products of the newly invented settled agriculture, and trade in them. Notice how the first tokens were not pure numbers, but numbers of things or stuff, like fat-tailed sheep or oil.

Body-part counting in the remote valleys of the New Guinea highlands with little traditional agriculture or trade served another purpose. These groups had a gift-exchange culture. The number of pigs given had to be remembered, if a suitable reciprocation were to be made.

Although there is plenty of evidence that ancient Australians traded widely, and over long distances, they seem to have done this face to face. This meant no need for bills of lading and invoices, as in ancient Sumeria, and therefore no need for writing. Presumably, trade meant person-to-person bartering: I give you this, if you give me that. For this kind of transaction no counting words are needed. Although sign language was widely used for inter-tribal communication, they don't seem to have signs for numbers either,[29] and they don't seem to make tallies on sticks, bones or stones.

Also, ancient Australians were essentially hunter-gatherers and thus had no food surpluses to trade on a seasonal basis, again unlike the inhabitants of the Fertile Crescent or New Guinea. But this is mere speculation. It is still a mystery as to why Australian languages do not have the usual complement of counting words, especially given that counting words up to five are so stable and so ancient in other language families. Could it be that the earliest speakers of Proto-Pama-Nyungan actually had these counting words, but they disappeared through disuse?

We have seen that the earliest writing systems were used to record the results of counting. There are also ancient non-writing systems, such as Sumerian tokens and Inca quipus, that recorded numerical information. Prehistoric members of genus *Homo* – *sapiens*, *neanderthalensis* and possibly *erectus* – recorded counts on bones, stones and cave walls. Very early humans almost certainly possessed counting words. This evidence strongly suggests that we and our ancestors could count and could express the results of counting because we already possessed concepts of number and counting, which in turn supports the claim that our ancestors inherited a counting mechanism.

CHAPTER 4

CAN APES AND MONKEYS COUNT?

Our ancestors from the genus *homo* not only counted and calculated but also recorded their counts on bones, stones and cave walls, even though they didn't have writing – signs for words. They probably spoke and heard counting words a hundred thousand years ago. We now consider whether our more distant hominid ancestors, the great apes, and more distantly, monkeys, are able to count and calculate, and what they can count. If we have inherited a counting mechanism from our common ancestor, then this should be implemented in a brain network similar to our own in the parietal lobes. In fact a region in the parietal lobe of the modern macaque monkey, with a common ancestor 30 million years ago, implements counting and some calculation. Macaques are moderately good counters, but the primate champions are the great apes, whose line diverged from our own only 6 million years ago.

Indeed, one of the most remarkable stories in the mathematical abilities of any animal is Tetsuro Matsuzawa's work with the chimpanzee Ai and her son Ayumu, at Kyoto University Primate Research Institute (KUPRI) in Japan. Matsuzawa describes its genesis:

> The day was 30 November 1977. A one-year-old female chimpanzee arrived at KUPRI, Japan. She was wild-born in the Guinean Forest, which spreads across four countries in West Africa: Guinea, Sierra Leone, Liberia, and Côte d'Ivoire. This means that she was a *verus* chimpanzee (*Pan troglodytes verus*). The infant was purchased through an animal dealer. Importing wild-born chimpanzees was still legal at the time. In the 1970s, Japan imported more than 100 wild-born chimpanzees, mainly for biomedical research of hepatitis B. This infant chimpanzee was one of them. However, instead of being sent to a biomedical facility, she was sent to KUPRI where she was to become the first subject of an ape-language research project in the country. The chimpanzee was soon nicknamed 'Ai' (pronounced 'eye'). Ai means 'love' in Japanese, and is also one of the most popular girls' name in Japan. She was estimated to have been born in 1976; hence she was about one year old at the time of her arrival.[1]

Around this time, there was great scientific excitement at the possibility of some kind of linguistic communication with great apes. Since their vocal apparatus is very different from ours, other channels were explored: American Sign Language (the sign language of the deaf) was used with the chimpanzees Washoe[2] and Nim Chimpsky, and with the gorilla Koko. Duane Rumbaugh at Georgia State University used a different method with the chimpanzee Lana: she was required to respond to 'lexigrams', abstract signs on a keyboard.

Rather than explore language abilities as such, 'what [Matsuzawa] truly aspired to do was to explore the perceptual world of chimpanzees through clearly defined visual symbols . . . How do chimpanzees perceive this world? Do they perceive it like we do?' Like Rumbaugh, he used lexigrams – black and white patterns rather similar to Japanese kanji characters – controlled by a computer to obtain objective, precise and detailed records of what was done and how the chimpanzee behaved. As well as these lexigrams, he investigated whether Ai could recognize and remember the twenty-six letters of the alphabet and the digits 0 to 9, using a match-to-sample method (see Chapter 1). Using this method and testing the ability to discriminate different stimuli, Matsuzawa and his colleagues tested short-term memory, sequence learning, perception of biological motion, colour vision and object recognition, face recognition and even visual illusions.

Number training began when Ai was about five years old. She had already learnt to use lexigrams corresponding to objects and colour names. Ai first learned symbols for eleven colours, e.g. ◇, "red", by pressing a key with the symbol on it. She then learned symbols for fourteen objects such as pencils ●, shoes, balls and spoons, by pressing the appropriate key. Ai was well able to do this, and this meant that Matsuzawa was able to train her on the numerosity of one type of object, say red pencils, and then introduce a new object or new colour, to see if she could still give the number of the new objects. In a separate test, she was asked to key in the number, the colour and the object in any order she chose, for example, *red, pencil, six*. As it turned out, number was always the last key pressed.

Notice that to match the numerosity of the set of objects to a digit, Ai must have a mental representation of the meaning of the digits that is independent of the nature of the objects in the set – that is, her representation of numerosity is abstract, at least relatively so. What strikes me as particularly extraordinary is the way in which five-year-old Ai can remember and use correctly eleven colour symbols, fourteen object symbols and six numerosities, and produce them correctly as required.

The great primatologist Jane Goodall was similarly impressed with Ai:

> When I first saw her she was in her enclosure with other chimpanzees. We made eye contact, and I gave the soft panting grunts that chimpanzees utter when they greet each other. She did not reply. An hour later I was crouched, looking through a small pane of glass, so that I could watch her at her computer. Matsuzawa warned me: 'She hates to make a mistake, and especially if a stranger is watching. She will bristle up and charge towards you and hit the glass window. But don't worry – it's bulletproof glass!' (Foreword to *Cognitive Development in Chimpanzees* (2006), edited by Matsuzawa et al.)

Matsuzawa and his colleague, Masaki Tomonaga, designed an experiment in which a set of dots is presented randomly arranged and different each time on one screen. On an adjacent touch screen were the digits, also randomly arranged and different each time, and Ai was required to touch the digit corresponding to the number dots. There were two experimental conditions. In the first, the dots were exposed until Ai touched a digit. In the second, the dots were exposed for 100 msecs and then masked with an abstract pattern so that they were no longer visible.

In the brief exposure condition, the chimp and the four humans performed with similar accuracy, but the chimp was much faster with more than five dots. In the unlimited condition, there was an increase in reaction time with the number of dots for both humans and Ai, although Ai was faster especially for nine dots, the most dots presented.[3]

Jane Goodall had observed Ai on another task, remembering the sequence of numbers briefly presented:

> One incident illustrates how these qualities [of concentration] enhance her success. Ai was working at a difficult task that involved memorizing a sequence of numbers on one computer screen so that she could replicate it on a second screen. A film crew was present, as well as myself. Ai, who is used to peace and quiet whilst she works, began to lose her concentration as first one and then another member of the team moved to get a better view, often bumping into the cage. She began to make mistakes – and after a few minutes her hair began to bristle. I was sure she was about to vent her frustration in a stamping display. Instead, she suddenly stopped working altogether, her hair sleeked, she sat very still and seemed to be staring at a point midway between the two screens. For at least thirty seconds, and maybe longer, she remained motionless. Then she started to work again. For the remainder of the session she paid no further attention to her noisy human observers. It was exactly as though she had decided that she must either give up or else pull herself together

and get on with the job! At any rate, whatever that pause meant, she made no further mistakes! (Foreword to *Cognitive Development in Chimpanzees* edited by Tetsuro Matsuzawa et al.)

Ai had been extensively trained, but what about her apprentice, her son Ayumu? 'I watched Ayumu working with his computer,' wrote Goodall. 'Like Ai, he seemed to have great concentration and loved to press the right panels for a small reward.'

The way Matsuzawa works is that a chimpanzee can *choose* to come into the lab to do a cognitive task. There is no compulsion, and there is no additional food reward for doing so. The chimps seem to enjoy the tasks. Also when an adult female chooses to participate in an experiment, she can bring her offspring. In the wild, the young may stay with the mother, and even suckle, up to five years old. When Ai brought in Ayumu, a touch-screen computer was provided for him so he could observe and imitate his mother if he so chose.

[He] began to learn Arabic numerals . . . at the age of 4. Before starting to practise, he had been observing his mother's performance on the computer since his birth. When his turn came, he first began by touching the numeral 1 followed by 2; this happened in April 2004. Later he learned to touch 1-2-3, then 1-2-3-4, and so on. Gradually, he succeeded in touching all numerals for 1 through 9 just like his mother. Ayumu then proceeded to the next stage: memorizing the numerals. Imagine that five numerals appear on the monitor. When Ayumu touches the first numeral, the other numerals turn into white rectangles . . . yet he is able to touch the rectangles in the correct [numerical] order. For this task, Ayumu has to memorize the numerals and their respective positions before he makes the first touch. Ayumu's performance in memorizing five numerals at a glance now exceeds that of his mother, and also that of human adults.

In this study, the interval between the display of the digits and the mask – how long the viewer has to take in the information – was varied. With a 650 msecs interval, humans are as good as Ayumu, though his mother, Ai, is not. With briefer intervals, Ayumu is better than humans.[4]

These studies show that in the lab chimpanzees can learn to exercise their numerical abilities to a very high degree. This suggests, but does not demonstrate, that their brains have inherited a 'number module' that enables them to count, learn the order of numerical magnitudes, and carry out simple calculations.

If they are possessed of this module, it is perhaps not that surprising that chimpanzees should be good at human tasks. Of all the animal species, the great apes – chimpanzees, gorillas, bonobos and orang-utans – are the most similar to us. Of these, the most similar are chimpanzees.

Chimps are members of the same family (*hominidae*) as humans (*Homo sapiens*), and their line split from our own only 6 million years ago – not much in the history of life on earth, 1 per cent of the history of vertebrates. Their genome is less than 2 per cent different from our own. Their brains are smaller, but not that much smaller, than our own: 384 grams vs 1350 grams in the average human, 28 billion neurons vs 86 billion neurons, with a very similar architecture.

Chimps' impressive cognitive capacities are not surprising. In the wild they will need a constantly updated cognitive map to locate fruiting trees, to work out when the fruits will be ripe enough to eat and, indeed, to remember whether fruit on a particular tree has already been foraged. They also need to know which part of a plant can be eaten and which can't.[5]

One lovely study in the 1970s by the American primatologist, Emil Menzel (1929–2012), demonstrated young chimps' ability to remember where the best food can be found.[6] Here's his experiment. One experimenter carried a test chimpanzee around the field, and a second experimenter hid one piece of fruit in each of eighteen randomly selected locations. The chimp could see all this, but couldn't respond in the normal way by approaching the food. Menzel's team repeated this on sixteen days with different hiding places each time. On average, the four test chimps found 12.5 out of the eighteen hidden foods, while two control chimps who had not seen where the food was hidden and just searched around the field found less than one on average. Menzel reports that 'Usually, the test animal ran unerringly and in a direct line to the exact clump of grass or leaves, tree stump, or hole in the ground where a hidden food lay, grabbed the food, stopped briefly to eat, and then ran directly to the next place, no matter how distant or obscured by visual barriers that place was.'

In a second experiment, instead of all fruit, half of the locations hid vegetables, which are less preferred by the chimps. Now the chimps went straight for the remembered locations of the fruit and ignored the vegetables. So the chimps had this extraordinary memory for eighteen locations that were briefly seen, and what was in them. But even more extraordinary was the fact that test chimps did not follow the experimenter's path, the path they could have remembered, but rather took the optimal path, the one which required the shortest distance – that is, the chimps had solved the 'travelling salesman problem' of finding the best route. This requires a mental map and calculation of a high order, but it is not yet known how the chimps do this.

Like us, in particular like our Stone Age human ancestors, chimps are territorial, social, live in small groups ('communities'), play, raise their young maternally and collectively, mourn the dead, attack other communities, and use and make tools. Since Jane Goodall reported detailed descriptions of chimps fishing for termites with a grass stalk, tool use has been the focus of intensive investigations of wild chimpanzee study sites across Africa.

They also communicate vocally and with facial expressions, and are omnivorous frugivores (that is, they will eat anything, including meat and insects, but prefer fruit if they can get it). They also cooperate, have ranks (the 'alpha male'), recognize other community members, and are capable of deception. Like us, they are active social learners. Infants learn from their mothers about troop practices, such as how to crack a hard-shelled palm nut using a stone anvil and a stone hammer. Typically, this is a kind of master–apprentice learning, where the master (mother) doesn't teach exactly, but enables the youngster to closely observe what she is doing, as did Ayumu.

In another chimp colony, Taï National Park, Ivory Coast, Christophe Boesch, of the Max Planck Institute for Evolutionary Anthropology, observed mothers actively teaching their young. Here is a description of one mother teaching her child:

On 22 February 1987, Salomé was cracking nuts of the very hard *Panda* species. Sartre, 6, took 17 of the 18 nuts she opened. Then, his mother watching, he took her stone hammer and tried to crack the nuts by himself. These nuts are tricky to open as they consist of three kernels separately embedded in a hard wooden shell, and the partly opened nut has to be replaced precisely each time to gain access to the different kernels. After successfully opening a nut, Sartre replaced it haphazardly on the anvil in order to attempt access to the second kernel. But before he pounded it, Salomé took it in her hand, cleaned the anvil, and replaced the piece carefully in the correct position. Then, with Salomé observing him, Sartre successfully opened it and ate the second kernel. Here, the mother demonstrated the correct positioning of the nut, although the infant may well have succeeded in opening it independently eventually.[7]

Different populations, especially widely separated populations, will, just like us, have different *cultures* – for example, different tool use. Only West African chimps (*Pan troglodytes verus*) do this. Central and East African chimps do not. Leaf folding is used to drink water, but the precise techniques vary in different populations.[8] Perhaps active teaching is also specific to some chimpanzee cultures. Like humans, there seems to be a sensitive period for acquiring certain skills. They can learn to crack nuts by the age of four to five (minimum three, maximum seven), while beyond that period they seem to have difficulty doing so.

Other great apes have similar numerical capacities to chimpanzees. One study compared chimpanzees (*Pan troglodytes*), bonobos (*Pan paniscus*), gorillas (*Gorilla gorilla*) and orang-utans (*Pongo pygmaeus*) on a task where they had to select one of two dishes with more food pellets from 0 to 10 pellets per dish. They only got the food after they had made the choice. This was true whether they could see the pellets in the dishes or after the dishes had been covered and they had to remember the number. There was no difference among the species in their performance on these two tasks – seeing and remembering.[9]

One possibility is that these subjects weren't counting at all, but just looking at the amount of food in each dish, or remembering the amount in each dish. Suppose, instead, the apes really did have to count and remember. For this task the subject – the ape – saw pellets drop one by one into cup A and then pellets drop one by one into cup B so that the content of the cups was not visible. He or she now had to choose a cup and would be more rewarded if the choice was the cup with more food pellets. Provided that neither cup had more than six pellets, all the apes apart from the bonobos managed to choose the cup with more pellets. If the comparisons involved numbers between six and ten, then none of the apes succeeded.[9] There seems to be, therefore, an upper limit to the number of items an ape can hold in memory in order to do a comparison task.

Chimpanzees, like us, coordinate their behaviour to hunt cooperatively; they also form coalitions to guard potential mates and patrol their territories. You may have seen videos of dominant males acting as guards when the community has to cross a road. What about counting? Can they count cooperatively? Here is another neat experiment from Matsuzawa's lab.

Consider the task on p. 125 where Ai and Ayumu had to touch the digits in numerical order. Could they cooperate so that Ai touched 1 and then Ayumu touched 2, Ai 3, and so on? Three mother–child pairs were used in the experiment. All were familiar with the sequence of digits.[10]

It turned out that all the pairs quickly learned the task already with minimal trial-and-error corrections and to almost perfect accuracy. So, not only could all these chimpanzees learn the digits from 1 to 9 in the correct order, they could cooperate to touch each of the digits in their correct numerical order. No problem.

These are extraordinary numerical skills for an ape to possess, but it is worth asking how these capacities are used in the wild. This is much more difficult to find out than by carefully controlled experiments in the lab. Is it possible to observe cooperative number use in the wild? One example comes from the work of Christophe Boesch. He was studying chimpanzee communities in the dense Taï tropical rainforest, where visibility rarely exceeds 20 metres. Nevertheless, the chimpanzees typically forage in parties of seven to twelve, but remain in contact with their community of about eighty individuals. They all move in a set direction for hours through this dense forest, in which one party cannot see the others. To do this they have to communicate auditorily. One way is to 'pant-hoot', a loud recognizable call, and the other is by 'drumming' on the highly resonant buttress roots of trees with hands or feet. This can be heard a kilometre or more away.

What Boesch observed forty years ago in one troop was quite remarkable. Brutus, the dominant male, drummed on the tree roots. When Brutus drummed twice on the same tree, everyone stopped to rest for sixty minutes before continuing their activities. 'Once I heard Brutus drum 4 times on the same tree, after which the community stopped for 2 hours 16 minutes. One example is not enough to draw any conclusions, but suggests that the number of bursts of drumming might indicate the length of the resting.' What is more, when Brutus drummed once on one tree and then twice on another tree, this would indicate a change in the direction of travel in the direction between the first and second trees.[11] This, as Boesch rightly points out, is symbolic communication, where one element of the symbolic system is numerical.

This numerical drumming, it turns out, was not unique to Brutus, or even to the Taï Forest. It was also observed by Jane Goodall in the Gombe Stream National Park, Tanzania, and in the Kibale National Park, Uganda. And in the Taï Forest, Boesch's team collected tape-recorded data from Brutus and another five adult males. Admittedly, the number used is limited to just two, though with one example of four, which was perhaps two twice.

Numerical assessment in the wild also occurs when two groups of chimpanzees are in conflict. As Goodall first observed in Gombe, chimps are territorial, with well-defined home ranges, and males will attack other males that invade their territory. It turned out that the five lethal attacks she observed involved parties with at least three adult males either attacking one individual from the competing community, or outnumbering the competitor. Were these killers comparing the number of themselves with the number of the competitors, and how can one tell?

The solution comes in a 'playback' experiment, which can be used with creatures too big, too strong or too dangerous to study in other ways. For example, it was inspired by Karen McComb's study of territorial defence by Serengeti lions (see Chapter 5). If a community of chimps *hears* nearby pant-hooting recognizable as being a stranger, possibly invading its territory, they have to decide what to do. Males may attack and try to kill a male invader, and they are more likely to be successful if they outnumber the invaders. In this case, the outnumbered party is likely to run away, and live to fight another day.

In the Kibale National Park of Uganda, Michael Wilson, Marc Hauser and Richard Wrangham of Harvard University tested whether the defenders really do make a numerical assessment of the number of invaders before attacking. They set up a loudspeaker playing pant-hoots from single foreign males on the edge of the Kanyawara community territory.[12]

The results were striking and straightforward. Three defenders or more would almost certainly counter-call, that is, make their own aggressive pant-hoot, and the probability of attacking the 'invader' increased with the number of defenders until eight defenders would almost certainly launch an approach to the loudspeaker.

Is this really numerical assessment? It has a degree of abstraction in that the defenders are comparing auditory information – the loudspeaker's pant-hoot – with information from other modalities – vision, for example – as the defenders look around to see how many of them there are. However, it's a case of n vs 1, where n can be from 1 to about 9. It would have been more interesting, I think, to see what would happen if the number of invaders also varied. Would it show the characteristic Weber fraction – the effects of proportional difference? This is what McComb and her colleagues did in a playback experiment with prides of lions, as we will see in Chapter 5.

There is some lab evidence that chimps can carry out simple calculations in the lab, as well as the more complex real-life calculations in the travelling salesman problem. For example, Sarah Boysen trained Sheba to associate digits with sets of objects, for example '2' with ■■, up to 5. Sheba was allowed to explore an area in which two digits were displayed, and then select one of five digits to get a reward. She soon learned to select the sum of the two digits up to 4 (1 comprised addends 1 + 0; 2: 0 + 2 and 1 + 1; 3: 0 + 3 and 1 + 2; 4: 1 + 3 and 2 + 2).[13] This, as Boysen notes, is very different from the way children learn about numbers (see Chapter 2). Even Ai's initial performance is different from our own. When learning a new digit, e.g. 3, she did not generalize from her knowledge of the meanings of 1 and 2 that 3 must mean a new numerosity, whereas human children after a while spontaneously generalize to the next number words and numerosities. However, it is worth noting that children are learning spoken words first, and in a range of contexts, just as they learn the meanings of other words they hear, so they start with an advantage over chimps, which lack language.

Monkeys

Monkeys are more distantly related to us than the great apes, with our last common ancestor 30 million years ago, compared with the apes at about 6 million years. Monkeys also have much smaller brains than the great apes. For example, the brain of the rhesus macaque (*Macaca mulatta*), which is frequently used as a model for human abilities, as we will see, weighs about 96 grams and contains 6.4 billion neurons, compared with the chimp's 384 grams and 28 billion neurons. Of course, size isn't everything, and although the structure looks similar, the monkey brain is not just a smaller version of the human or chimp brain. The prefrontal cortex, critical in human cognition, and probably in chimps too, is relatively smaller in the macaque. And as we will see later, the tiny brains of fish and even insects can count. What a creature can count and how far it can go is the crucial question.

The first study to show decisively the numerical abilities of monkeys was carried out by a brilliant student at Columbia University in New York, Elizabeth Brannon, and her supervisor Herb Terrace.[14] We will be hearing a lot more from Elizabeth in this chapter. Their study was a very stringent test of numerical ability. After training to select the larger numerosity from pairs of sets of objects from one object to four objects, the macaque monkeys Rosencrantz and Macduff were offered two new pairs, and what counted was their response on the very presentation of the new pairs. Some of the new pairs used sets of objects that were familiar from the training, and some were novel objects. Rosencrantz and Macduff were rewarded for selecting the larger on the training trials. This is an example of 'instrumental conditioning' and Terrace has been a student at Harvard of the high priest of this paradigm, Fred Skinner. To be sure they were responding to numerosity rather than other visual features, very extensive controls were used. Figure 1 (overleaf) shows some of the stimulus sets that were used.

The monkeys had learned numerical ordering whether both sets were of familiar objects, or when one had a set of novel objects or indeed both sets comprised novel objects.

Monkeys are not confined to counting things they see or hear; they can also count things they do. (We'll see much more on action counting in Chapter 5.) Jun Tanji and his colleagues at Tohoku University in Japan required monkeys (*Macaca fuscata*) to push a lever five times, wait 1.4–7.5 seconds for a signal, turn the lever five times, and then repeat the cycle. It took a lot of training, but after ten months they were very accurate, almost always doing the correct action at least four times.[15]

Animal cognition expert Hank Davis, from the University of Guelph in Canada, believes that although animals *can* use

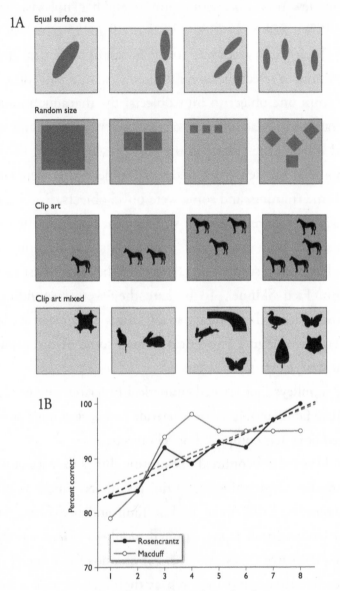

Figure 1. 1A: Examples of stimuli used in the first numerosity ordering experiment in monkeys. 1B: results of the test trials where two new sets were presented. Weber's Law is followed: the larger the difference between the two sets, the more accurately the larger set was selected.[14]

numerical information, they only do so as a 'last resort', when other quantitative measures – area, volume, time – or non-quantitative properties – colour, shape, odour – can't be used to make the decision.[16,17] A very beautiful experiment by Elizabeth Brannon and Jessica Cantlon, then at Duke University in North Carolina, suggests that actually macaques make numerical assessments as a *first resort* when making a choice.[18] To test this, they used a match-to-sample task (see Chapter 1) which offered the monkeys the option of matching on the basis of number or of other potentially relevant dimensions – colour, surface area or shape. So, for example, as shown in Figure 2A (overleaf), the monkeys were trained and rewarded for choosing two discs, so they could have been responding to discness or twoness. In a 'probe trial', where two discs were not on offer, they could choose on the basis either of the number – two daggers – or on the basis of shape – four dots.

Similarly, as shown in Figure 2B (also overleaf), they were trained to choose one large square, and on the probe trial, they could choose on the basis of number – one – or area.

The essential result was that the monkeys indeed use number as a first resort, but when the ratio is too difficult – 25 per cent or less – or the numerosity is too large – eight or more objects – they use the other dimension to make their choice.

In fact, the ability of macaques to make numerosity discriminations is similar to, though not quite as good as, humans. Both species show a striking ratio effect.[19]

Yet another parallel between monkeys and humans discovered by Brannon's lab concerns arithmetic and was reported in the intriguingly titled paper, 'Basic math in monkeys and college students'.[20] Of course we would not expect monkeys to be able to

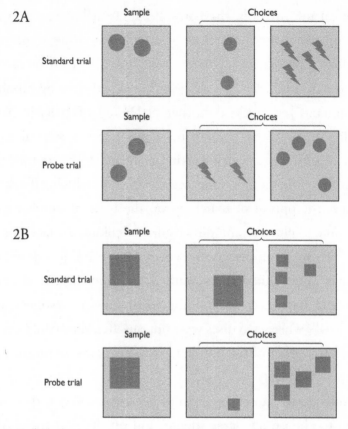

Figure 2. Match-to-sample tasks. Monkeys are trained on the 'standard trial' and then have to make a choice on the 'probe trial', which does not include the stimulus on which the monkey was trained. On the probe trial 2A, the monkey can choose on the basis of numerosity (two) or shape, and in 2B can choose on the basis of numerosity (one) or area. In both cases monkeys more often chose on the basis of numerosity than the other property.[18]

produce the correct answer to a problem presented as '3 + 2 = ?' (unless they are specifically trained to use numerals; see below). The arithmetical problems need to be presented in a different way. The monkey or human would see a panel with, say, three dots, then half a second later a panel with four dots, and half a second later a panel with two smaller panels, one with seven dots (correct) and one with, say, four dots (incorrect). The subject's task was to touch the subpanel with the correct number. In the initial training, the addends were 1 + 1, 2 + 2 and 4 + 4.

The later trials tested all possible addends of the sums 2, 4, 8, 12 and 16. For example, when 8 was the sum, the addends could be 1 + 7, 2 + 6, 3 + 5, 4 + 4, 5 + 3, 6 + 2 or 7 + 1. Both the two monkeys, Boxer and Feinstein, and fourteen college students could do this task, and both showed similar patterns. Accuracy depended on the ratio of the two choices (the Weber fraction yet again), and both groups showed a 'size effect'. That is, accuracy depended on the size of the sum, with larger sums being less accurate. Cantlon and Brannon made sure that the monkeys (and the humans) were not using the total surface area of the dots to make the choice, and also that the choices were balanced so that simply picking the larger set would not be a good strategy. Of course, there was an important difference between the monkeys and the college students. The humans could do the task right away, but it took at least 500 training trials for the monkeys to get to above-chance performance, and they each completed five thousand training trials before they started on the later trials. (The amount of training is one of the reasons why monkey studies are very hard work for both the subjects and the experimenters.)[20]

It is possible to teach monkeys symbols. In a study lasting several years, three macaques have been taught the symbols 0, 1, 2, 3, 4, 5, 6, 7, 8, 9 for quantities 0 to 9 delivered as the number of drops of juice, and the letters X Y W C H U T F K L N R M E A J representing 10–25 drops. The Harvard neuroscientist Margaret Livingstone and her colleagues based a really ingenious experiment on the monkeys' symbolic skills. They tested their ability to add two numbers.

The basic set-up was like this. On the left or right side of the screen was a number, say 6, and on the other side two numbers, say 4 and 3. The sum of 4 and 3 is greater than 6 so the reward of drops would be greater if they chose the sum rather than the 'singleton'. In the initial training, dots were used rather than symbols, but of course the dots on the right might not be added but only counted. The symbol condition does not require counting. However, the monkeys had a lot of experience of these symbols and their relationship to the number of reward drops, so they might have been using some kind of association rather than real addition. To get round this possibility, the experimenters taught the monkeys a whole new series of symbols made up of squares to represent the numbers 0 to 25. Here are some examples: ⌐⊦⌐⌐⌐⌐⌐. For example, a monkey subject could select the sum of two symbols (9 + 13 = 22 drops) of the screen rather than the singleton symbol worth 19 drops.[21]

It is possible to model how the monkeys represent the value of the sum in relation to the subjective value of the addends in the dot, digit and novel symbol conditions. Roughly speaking, the mental representation of the sum is indeed the linear sum of the mental representation of the addends for all three conditions. However, small numbers are sometimes undervalued and large numbers sometimes overvalued, but this depends on all the numbers in the display.

These are very beautiful lab studies, but do monkeys use this numerical ability spontaneously and without training?

Doing what comes naturally

It is possible to find out using a 'field experiment'. One early study mimics natural foraging by macaques in a free-ranging situation on the island of Cayo Santiago, Puerto Rico. The monkeys live in a large protected area, and if they so choose, they can participate in experiments. This is how numerical abilities were assessed by animal behaviour experts Marc and Lilan Hauser, and Susan Carey, a psychologist who is, as we have seen, a major figure in the development of numerical abilities in human infants and children:

> Two researchers placed themselves 2 m apart and 5–10 m away from the test subject. Each researcher then showed the monkey a distinctively coloured, opaque box and emphasized that it was empty by tipping it sideways and placing an open hand inside; each researcher then placed the box on the ground in front of his/her feet. One researcher then placed one or more [apple slices] into the box, making sure that the monkey watched the events. When the objects were in place, the researcher stood up and looked down. Subsequently, the other researcher placed one or more [apple slices] into his/her box and then stood up and looked down. Having completed these events, both researchers then turned and walked away in opposite directions.[22]

There was no training, so the observed behaviour was what the monkeys would do naturally, as if they were seeing two branches with different numbers of ripe fruit on them. So do the monkeys then select the box with more apple slices? To do so means that they would have to do three things: count (or add) the slices in box one, count (or add) the slices in box two and compare the results. In terms of the simple accumulator model in Chapter 1, there is a 'reference memory' for box one and a 'working memory' for box two, so that the comparison process needs only to compare the quantities in the two memories. In the wild, monkeys succeed by gathering the most fruit with the least effort.

Here 15 monkeys were tested, and they had no trouble selecting 2 slices vs 1 slice, 3 vs 2 slices, 4 vs 3 slices or 5 vs 3 slices, but they failed when the proportional difference (Weber fraction) was small – 5 vs 4 (25 per cent) – or where one of the numbers was larger than 5.

Now it remains possible that monkeys are not really using counting or addition, but just keeping track of the total quantity of food in each bucket, though this strategy would require them to add or summate the quantities of each slice, and this seems to a human like me a more difficult task.

There's another experiment by Brannon and her team that needed no training at all, again stressing what comes naturally to the macaque. In fact, it demonstrates the monkey's capacity to use an abstract representation of numerosities.

The macaque subject saw one to three videos of other macaques vocalizing. For example, there could be three vocalizers on one monitor and two on the other. At the same time they would hear either two vocalizations or three vocalizations. The monkey (*Macaca mulata*) hears either two monkey or three monkey vocalizations. It looks longer at the video where the number of vocalizers heard matches the number of vocalizers seen.[23]

There was no task; Brannon and her colleagues simply recorded the time the subject monkey spent looking at the three or the two monkey videos. It turned out that they looked longer at the monitor that matched the number of vocalizers to the number of vocalizations heard – that is, they matched the numerosity of the vision to the numerosity of the audition more often than would be expected by chance.

One of the remarkable features of the Brannon lab is that she, her students and colleagues work on two notoriously difficult subject groups, monkeys and human babies, often using the same experimental paradigm with both groups. I think the lab is the only one that rises to this challenge: others work on one or the other. So naturally, Brannon also did the parallel experiment with babies. Seven-month-old infants heard either two human voices or three human voices. They look longer at the video where the number of voices heard matches the number of speakers seen.[24]

In more naturalistic settings with free-ranging monkeys in Cayo Santiago in Puerto Rico, Marc Hauser and two colleagues from Yale University replicated a study that Karen Wynn carried out with human infants (see Chapter 2). It showed that the monkeys, like small young humans, could carry out additions.[25] The logic is simple: will the monkeys respond differently when the outcome of an addition is impossible? Here is one example. The monkey is shown three lemons, a preferred food, which are then covered by a screen. The monkey is then shown one lemon going behind the screen. The screen is then removed to reveal a number of lemons. The reveal can show four lemons, the correct sum of 3 + 1, or eight lemons, an impossible sum of 3 + 1. Will the monkey respond differently to the two reveals, for example by looking longer at the impossible outcome? Wynn's infants did look longer at impossible outcomes. Would the monkeys do the same? Of course, eight lemons are more interesting than four, so the experiment must control for this. In one condition, the monkey saw 4 + 4 = 4 or 4 + 4 = 8. If it was looking longer at the larger quantity of lemons, then it should look longer at the eight, but if it had what I have called an 'arithmetical expectation' then it should look longer at the four, which indeed it did.

There is also the possibility that monkeys are responding to the total amount of lemonness, and that can also be controlled for by using large, medium or small lemons to see if the monkeys are using numerosity only as a last resort in this situation. Notice that the monkeys are not being trained: they only have one trial and so what is measured is what comes naturally to the monkey.

The critical variable, as in many other situations, is the Weber fraction, the proportional difference between the two outcomes. This is a good indicator that the monkey was indeed using a numerical process. So for example, it looked for the same amount of time when the outcome for 2 + 2 was 6 as when it was 4, the difficult 2:3 ratio. Again we see the similarities between untrained monkeys and untrained infants.

Monkey mafia and monkey business

At the Uluwatu temple in Bali, Indonesia, free-ranging long-tailed macaques (*Macaca fascicularis*) mean business. They rob tourists, but are willing to barter the stolen goods for food. This is learned behaviour. Older monkeys do it more than younger monkeys, some groups at Uluwatu do it more than others, and other monkeys in sites with many tourists don't do it at all. The monkeys seem to prefer to steal glasses, hats, shoes and cameras, since these are often not properly secured, and are easily taken.[26] But why would these temple monkeys want your camera? It has no intrinsic value for the monkey, but they have learned that these tokens can later be exchanged for objects that are of value to the monkey, namely food.

But will the objects that have most value to the tourist elicit the biggest food reward in the bartering? This is the question the ethologist Jean-Baptiste Leca at the University of Lethbridge in Canada and his colleagues asked.[27] Mobile phones, wallets and prescription glasses are among the high-value possessions the monkeys aim to steal. 'These monkeys have become experts at snatching them from absent-minded tourists who didn't listen to the temple staff's recommendations to keep all valuables inside zipped handbags firmly tied around their necks and backs,' said Leca in an interview.[28] Juveniles didn't seem to realize the value of the stolen items but older monkeys did, and targeted high-value items more than medium-value items, and medium more than low. Not only that, but when objects of different value were accessible on the same tourist, the older monkeys tended to select the most valuable. These clever monkeys, at least the most skilful of them, not only stole more valuable items, but had learned that these items could elicit a greater pay-off in the bartering process, and would reject a low offer and wait until more food or preferred food was offered. Less skilled juveniles and subadults would tend to accept the first offer even for high-value objects. They haven't yet learned the value of their stolen goods for the bartering process.

Leca and colleagues call the skilful behaviours 'value-based token selection' and 'robbing/bartering payoff maximization'. 'You wouldn't want anything to happen to your nice camera, would you?' It's a monkey protection racket by a macaque mafia.

There's certainly calculation in this behaviour, but does it involve counting or anything that's really numerical? The experienced monkey has to have in memory a subjective scale of value of different types of robbable objects. This scale could be just be an ordering. For example, camera > glasses > scarf. Or it could be that each object type is given a numerical value: camera = three pieces of fruit, glasses = two pieces of fruit, scarf = one piece of fruit. This could be directly exploited in the bartering process. Leca and colleagues consider a wide range of cognitive capacities needed for these behaviours:

> Preference transitivity, self-control, delay of gratification, action planning and calculated reciprocity . . . may facilitate or constrain an individual's ability to make optimal economic decisions. Even though these characteristics were not explicitly examined in this study, some of them will be the subject of our future observational and experimental investigations.

But not, alas, numerical abilities.

Baboons calculate with their brains
and vote with their feet

Baboons in laboratory-type settings show numerical abilities. In fact, in one very interesting study it is possible to show that they can do something that is logically equivalent to counting, again from Elizabeth Brannon's lab (see Figure 3).[29]

The monkeys were not differentially trained to choose the larger set: they simply got the number of peanuts in the bucket. They keep track of the number going into the second bucket, and when it reaches the same number as the first bucket they start to move towards it. That is, they are counting the numbers and then making a comparison before all the data is in, just like we might do.

a) Cache I is sequentially baited with 5 food items

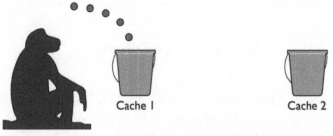

Monkeys typically sat in front of cache I while it was being b

b) Then cache 2 is sequentially baited

Monkeys typically moved to cache 2 when it was approximately equal to cache I

Figure 3. Example trial. Baboons were presented with a choice between two sequentially presented sets of peanuts. The monkeys frequently committed to a choice after the first bucket (cache) was completely baited but before the second cache was completely baited. The baboons indicated that they had reached an early decision by physically moving from the first set to the second set in the middle of a trial. This suggested that they were counting the second set and comparing this count with the numerosity of the first set.[29]

And like us, the probability of a switch from Bucket 1 to Bucket 2 increased as the number of peanuts going into Bucket 2 approached and then exceeded the number in Bucket 1. The other feature, as you might now come to expect, is Weber's Law all over again. The probability of preferring Bucket 2 depended on the ratio of the number of peanuts in the two buckets: the bigger the difference, the more likely the preference for the bucket with the more peanuts.

Voting with their feet

It turns out, in an extraordinary study of wild baboons, that they deploy their counting ability in the wild to decide in which direction to move. Baboons in many parts of Africa live in highly cohesive 'troops' of up to about a hundred males and females. The troop will move together when foraging or choosing a sleeping site. The question arises as to how the troop 'decides' where to go. Does the troop follow one of the dominant males? Does it follow the one who initiates the move in a particular direction? Previous research found no decisive evidence for either theory.

In order to solve this problem, 'an observational task of daunting dimensions' according to the primatologist Richard Byrne, the movements of one troop of olive baboons (*Papio anubis*) that lived in the Mpala Research Centre in Kenya were studied by an international team of baboon experts – Ariana Strandburg-Peshkin, Damien Farine, Ian Couzin and Margaret Crofoot.[30] To do this, the team took the extraordinary step of equipping 80 per cent of the adult and subadult members of one troop with custom-designed global positioning system (GPS) collars to record 'movement initiations'.

The team knew the sex and hierarchical status of all these baboons, so they could test whether indeed it was a simple case of 'follow my leader'. The experiment turned out to confirm previous studies that this was *not* what happens. It showed that the baboons made decisions about the direction of troop movement democratically, on the basis of the relative number of animals heading in different directions. So if A is heading north and B is heading west, for example, most of the troop would wait to see which of them has gathered the most followers and then join that group, let us say A, then the group around B would join the group around A. The team wondered whether it really was counting, or did the baboons use some other measure, such as the total mass of baboonness? Because they knew each individual, and their weight, they could actually calculate the mass of baboonness of each group. What is more, an individual baboon's decision follows Weber's Law: that is, when the proportional difference between the group around A and the group around B is large, then the baboon will choose A more quickly than if the difference is small.

Monkey brains

Another way of looking at the relationship between other primates and ourselves is to ask whether the brain system that carries out numerical tasks is similar. If it is, this would be a strong argument that the numerical capacities of us and them are derived from a common ancestor.

We saw in Chapter 2 that there is an area in humans that carries out 'summation coding', an accumulator, located in the superior parietal cortex close to the intraparietal sulcus in the left and right hemispheres.[31] We also know that the intraparietal sulcus supports numerosity judgements in humans, whether the objects are presented simultaneously or sequentially[32] or whether sequential objects are visual (squares) or sounds (beeps).[33] We also know that these areas are critical to numerical abilities since if they are damaged, simple numerical tasks, such as assessing or comparing the numerosity of sets, become difficult or impossible (see Chapter 2).

A study from Duke University by Jamie Roitman, with Elizabeth Brannon and Michael Platt, identifies neurons in a roughly homologous region of monkey cortex (the lateral intraparietal cortex) that acts like an accumulator in that the more dots the monkey sees, the more these neurons fire.[34]

Monkey studies can locate the number regions of the brain far more accurately than human or indeed chimpanzee studies. Human studies rely on neuroimaging techniques such as functional magnetic resonance imaging (fMRI) that can, at best, pick out a region containing several 'voxels' (see p. 71), each of which has at least 1 million neurons and many millions of connections among them. Because the nature of the technology depends on the uptake of oxygen in the neurons, there will be a delay, often of several seconds, before the activity becomes detectable. There are other techniques that reduce the time delay but have a much coarser spatial resolution. In cases of intractable epilepsy it is sometimes possible to expose the brain surface during surgery and measure the effects on perhaps a few hundred neurons.[35] With monkeys, a standard procedure, used by Roitman and her colleagues, is to insert an electrode probe into the anaesthetized monkey's brain stereotactically using 3D imaging so that the exact location of the probe that will be recording from perhaps just a single neuron, and because this depends on an electrical signal rather than oxygen uptake, the response – the firing rate of the neuron – can be recorded immediately. The more active the neuron, the greater the firing rate. In the alert monkey, then, it will be possible to tell whether, for example, the firing rate increases monotonically with the number of objects seen, as in the Roitman et al. study.

The question then arises as to what happens to the contents of the accumulator in the monkey lateral intraparietal cortex. There has to be a 'transfer function' that maps from an accumulator level to a stable representation of numerosity – that is, every time the level is about here, that means something like 'fourness', and about here, 'fiveness', and so on. It's rather like a thermometer where the mercury level is calibrated with numerical temperatures, and indeed this system is often referred to as a 'thermometer' coding or 'labelled line' coding. In a proposal by Stanislas Dehaene and Jean-Pierre Changeux, the accumulator level is mapped on to a kind of labelled mental number line such that a specific neuron responds most strongly to one specific numerosity, say fourness, though it will respond less strongly to neighbouring numerosities – threeness, fiveness, twoness, sixness and so on (see Chapter 1). Children learning verbal counting need to map the accumulator level to the counting words: this level means 'two', that level means 'three' and so on (see Chapter 2).

If the monkey's accumulator neurons are in the lateral intraparietal sulcus, where is the labelled line, and how does it work? The answer was discovered by the neuroscientist Andreas Nieder working in Earl Miller's lab at MIT, and then in his own lab at the University of Tübingen in Germany and by Jun Tanji and his colleagues at Tohoku University in Japan.

As we saw on p. 135, Tanji's monkeys learned to push a handle five times and then switch to turning the handle five times, and then repeat the cycle. Recordings from neurons from the left parietal lobe, in particular in part of the intraparietal sulcus, revealed that these cells fired when the monkey was counting the actions.

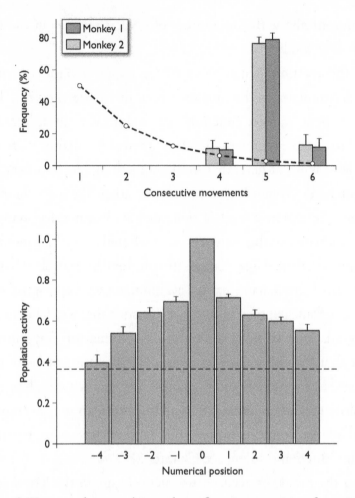

Figure 4. Two monkeys pushing a lever five times, waiting for a signal and then turning the lever five times. The upper panel shows that the vast majority of responses are accurate, and the errors ±1. The lower panel shows the activity of cells in the parietal lobe according to the number of actions, with peak activity on the correct number, with reduced activity for neighbouring numbers.[15]

That is, there was peak activity for cell A during the first push or turn; for another cell, the peak activity corresponded to the second push or turn; and so on (see Figure 4).

Nieder used a delayed match-to-sample method where the

monkey would see one, two, three, four or five blobs on a screen and after one second was then given a choice between a panel that looked different from the sample and another with a different numerosity. For example, the sample could be three blobs and the choices would be three and four blobs. Naturally, very careful controls were carried out to ensure that the choice was based on numerosity not any other visual features. In the first study, a trial started when the monkey grasped a lever. The task required the monkey to release the lever if the sample and test displays contained the same number of items and to continue holding it if they did not.

Nieder recorded from two regions in the monkey brain that were known in humans to be involved in number tasks: the frontal lobes and the parietal lobes. The critical evidence came from the recording of the second between the presentation of the sample stimulus and the presentation of the choice. The critical neurons turned out to be very near where Roitman's accumulator neurons were located. They were in the ventral intraparietal cortex, that is the bottom – or 'fundus' – of the intraparietal sulcus, close to the lateral intraparietal sulcus. These neurons, Nieder discovered, were tuned to particular numerosities; that is, they fired most when the monkey saw a particular numerosity. So, for example, the 'three neuron' fired most strongly when the monkey was remembering the three blobs just presented. These were not accumulators since their firing rate was not proportional to the numerosities they were tuned to.

That is, the 'three neuron' would fire not only to a set of three dots, but also to two, four and five blobs. So the tuning was not exact, but approximate, and thus Nieder and many others describe the underlying model as the 'approximate number system' (see

Chapter 1).[36] This creates a serious problem if we are trying to identify the brain mechanisms for counting. We humans don't count 'sort of three, but it could be two or four, maybe even one or five'. We count three, exactly. I have described how human children learn to calibrate the internal accumulator so that levels correspond to the counting words. This doesn't mean that children are always accurate in their count. They can make errors by missing an object or by double counting, and the accumulator may be a little fuzzy. Children have the words to help them learn to count accurately. Monkeys don't, normally. However, when they are taught numerical symbols, as we saw earlier, they can be remarkably accurate in representing quite large numerosities and calculating with them.

A critical feature of Nieder's study and that of Roitman et al. is that the critical brain region for numerosity processing is homologous with the human brain's key region. This is particularly striking when you compare the region we identified in the human brain with that identified by Nieder and his colleagues. We both found that the key region is not just in the parietal lobe, but in the bottom – the 'fundus' – of the groove in the cortex known as the intraparietal sulcus. This can't be a coincidence. It suggests that the numerosity processing mechanism in both species is inherited from a common ancestor.

The studies, published in the same year, both showed that these core regions responded to a numerosity whether presented all at once as an array of objects or one by one. And it is crucial to the abstract idea of a numerosity that it is independent of the mode of presentation as well as the visual features of the objects presented.

In a subsequent study, Nieder showed that some neurons in

the monkey intraparietal cortex responded only to numerosities, while other neurons in a similar delayed match-to-sample task responded only to continuous quantities, in this case the length of lines; and a few responded to both numerosity and length.[38]

As it happens, our study also looked at numerosity vs continuous quantity. In our study it was the relative areas of blue and green compared with the relative numerosity of blue and green squares, and we showed an activation for numerosities distinct from continuous quantity.

It's not quite true that specific neurons with numerical functions have not been identified in humans. However, this entails mapping the exposed brain with electrodes to identify the brain tissue requiring surgery. One example is where the patient has a glioma, a form of cancer, that needs to be surgically removed. One very important study by Carlo Semenza and his colleagues at the University of Padua and San Camillo Hospital in Venice was able to explore the parietal lobes of nine awake patients during surgery.[39] Their method was to use the electrodes to stimulate neurons which would temporarily prevent them working normally. In this way, they were able to show that there are several loci in both left and right parietal lobes and neighbouring regions that were involved in multiplication and/or addition.

In very severe and debilitating epilepsy that doesn't respond to pharmacological or other treatments, surgery is also required to remove the epileptic focus. Most epileptic foci are in the temporal lobes and Nieder and his colleagues located neurons in this region that behave similarly to those in monkey IPS when recorded during a simple calculation task. The temporal lobes, especially the left temporal lobe, are the seat of what is called 'semantic memory' where facts are stored, including word meanings and facts about the world, such as *banana is a fruit, Paris is the capital of France* and so on, so it is not surprising that there are representations of numerical facts both non-symbolic (dots, for example) and symbolic (digits) in this region.[38]

Nieder's work with monkey brains has also revealed that the frontal lobes are part of the number network, with neurons there representing numerical information more abstractly. We also know that human brains are connected up in a similar way, with the frontal lobes, especially the left frontal lobe forming a network with both left and right parietal lobes, so that activation in the parietal lobes sparks activation in the frontal lobes.[40] Learning new arithmetic in humans strengthens the connection from the left and right parietal lobes and left frontal lobe.[41] In fact, it is possible actually to trace the white matter fibres (the axons that convey information from one neuron to another) that carry this information during numerical tasks.[42]

Our nearest relatives, the apes and monkeys, naturally see the world in numerical terms without any help from us, but we can in the laboratory and in cunning field experiments reveal something of how and why they do it. Like us, their abilities are characterized by Weber's Law. We know from the ingenious experiments of Andreas Nieder and Jun Tanji that monkeys use homologous (equivalent) brain structures to humans to carry out numerical tasks. We can therefore be reasonably confident we have inherited the basics of our numerical abilities from the common ancestor of these primates and us some 30 million years ago.

CHAPTER 5

MAMMALS GREAT AND SMALL

Our closest non-human cousins, apes and monkeys, are able to count and carry out calculations. Our more distant cousins among other mammal species, it turns out, can and do count and calculate, from those with small brains, such as mice, to those with brains bigger than humans'. In the lab, rats and mice are extraordinary counters, but animals that are too big or too expensive to test in the lab reveal numerical abilities in the wild that can save them from injury or death, and support group advantage.

Deadly calculations in the wild

Imagine if you will that you are a female lion trying to defend your pride from invaders. Your territory is the savannah of the Serengeti National Park in Tanzania, with its high grass that can hide invaders from sight. What is more, these encounters typically take place during the hours of darkness or half-light. You can't see them, but you can hear lion roars approaching closer and closer. Which lions are they? Are they from your pride or are they invaders? You can distinguish the roars of the fathers of your cubs from the roars of unfamiliar males, some of which might well be infanticidal males. If the roars are unfamiliar, what are you going to do? Maybe the roars are from female intruders. That's another problem. Fights between lions can lead to serious injury or death.[1] Again, what are you going to do? Fight or flight? If you outnumber the intruders, they will probably back off – a bloodless win for your pride. If they outnumber you, they will attack, you will have to flee and the intruders may take possession of your territory, your females, your cubs. Getting the numbers right can be a matter of life and death.

Female lions (*Panthera leo*) are social and cooperative. They live in prides of up to eighteen females and 'advertise their joint ownership of a territory by scent marking and roaring. Inter-pride encounters sometimes result in intense chases which are more likely to be won by the larger group of females . . . but fighting entails a high risk of serious injury and is rarely observed.'[2] This means you have to assess how many intruders there are from their roars, and how many of you there are. Do you outnumber them?

One could just stake out a pride and wait to observe what happens when there are invaders. One could wait a very long time for a relevant event to occur. Instead of just sitting and waiting, Karen McComb, then at Cambridge University, and her colleagues Craig Packer and Anne Pusey of the University of Minnesota devised a way to test whether pride members indeed carried out this numerical assessment. McComb had developed a method she called 'playback' using loudspeakers. She had used this method to discover what kind of male roaring female red deer (*Cervus elaphus*) find most attractive. She manipulated pitch to see whether hinds preferred males with a deep roar, or the number of roars per 'bout', and played the recordings through loudspeakers near the hinds. It turned out that there was no preference for the low-pitch roaring, but there was a highly significant preference for males who made more roars per bout, though McComb does not, in this study, focus on the numerical abilities of the hinds.[3]

Using a similar technique with lion prides, she placed loudspeakers near the boundaries of the prides. She would play roars from one unfamiliar female 'intruder' or from three intruders. The number of defenders would vary. It was then possible to plot the likelihood of pride members attacking the 'intruders'. It turned out that three defenders are likely to approach the speaker with the intention of engaging a single intruder, but highly unlikely to approach three intruders. Six defenders will almost always approach three intruders. However, when there are cubs present, the defenders will always approach.

In a comparable study of male lions in the Serengeti, the pride males make a numerical assessment of themselves and the intruders before deciding to approach a loudspeaker, and will wait for other pride males to join in before approaching if the ratio between them and the intruders is unfavourable. However, the resident male will almost always approach the speaker.[1]

One really important issue from these studies is that these assessments depend on a numerical comparison between the number of roarers you can hear and the number of nearby members of your pride that you can see. That is, these are cross-modal comparisons, indicating that the selector identifies *lions* irrespective of the sensory modality, and updates the accumulator accordingly.

This brilliant innovation of using playback to test the use of numerical information in inter-group conflict has been adopted with other species. We saw in the previous chapter how it was used with chimpanzee troop conflicts, but it has also been used with spotted hyenas (*Crocuta crocuta*). Hyenas, like lions, live in 'fission-fusion' societies, where group size can vary dramatically, and with their sharp teeth and powerful jaws, inter-group conflicts can be lethal, and the larger group usually wins. Sarah Benson-Amram, another brilliant contributor to the Royal Society meeting on the origin of numerical abilities,[4] and colleagues at Michigan State University used playback to test numerical assessment in two study hyena clans in Maasai Mara National Reserve in Kenya with long-distance contact calls, known as whoops, produced by one, two or three unknown hyenas, or 'intruders'.[5]

> In fission-fusion societies, it would not be uncommon for a relatively small subgroup from one clan to encounter a relatively large subgroup from a neighboring clan. In these situations, the larger subgroup can attack the smaller subgroup at relatively low cost to themselves. Therefore, the greater variation in subgroup size in fission-fusion societies may lead to higher rates of intergroup aggression. Animals living in fission-fusion societies are likely under increased selection for the ability to assess numerical odds, or the ratio of number of defenders to number of intruders, before engaging in aggressive intergroup interactions.[4]

In a variation of the lion study, they presented the whoops sequentially rather than all at once in a chorus, and they ensured that the defenders heard the same number of whoops, whether from one, two or three 'intruders'. As with the lions, the hyenas' approach to the speakers depended on the ratio of defenders to intruders. Again, notice that the comparison is cross-modal, and hence abstract, because the intruders are only heard while the defenders can see each other. In carrying out the numerical comparison between themselves and intruders, like lions, the selector identifies hyenas irrespective of the sensory modality, and updates the accumulator accordingly.

Guinea pigs and lab rats

Small, inexpensive animals, rats and mice – the modern 'guinea pigs' – have been very extensively tested in the lab, but with little interest so far in their wild life – but these creatures have revealed exceptional numerical abilities.

I'll start with rats since many of the key ideas in animal counting come from these studies. B.F. Skinner (1904–1990) was a professor at Harvard and a key figure in behaviourism. His method of 'operant conditioning' – teaching animals by selectively rewarding desired behaviour – could achieve extraordinary feats. In 1958 a brilliant PhD student at Columbia, Francis Mechner, published part of his thesis in the *Journal of the Experimental Analysis of Behavior*, the leading journal of the Skinnerians, and this study followed Skinner's principles and methods, including a Skinner Box, and what Skinnerians called a 'fixed ratio schedule', where the animal is rewarded after a certain number of responses. Mechner invented a neat variant on the fixed ratio paradigm, in which a rat would be rewarded by pressing lever A in the Skinner Box a certain number of times, and then switch to lever B.[6] If the rat pressed A fewer times before switching to B, it would not be rewarded. He showed that the rat subject could learn to press lever A four, eight, twelve and sixteen times. In addition to this demonstration of numerical ability, the data showed two other interesting features. First, as the required number got larger, the proportion of errors became greater. This is known as 'scalar variability', as I mentioned in Chapter 1. Mathematically, the standard deviation of the responses divided by the mean, in this case the target number – the 'coefficient of variation' – is a constant. When the coefficient of variation is constant, the discriminability between, say, eight and twelve presses becomes a function of the ratio between the to-be-discriminated quantities and we're back to the familiar Weber's Law.

Second, the rat was more likely to err by pressing too many times than too few – it was rewarded if it pressed too many times, but not too few. I will return to what this means when I describe mouse experiments.

In 1983, Warren Meck and Russell Church, then at Brown University in Rhode Island, published a study that provided the first evidence for an accumulator mechanism that could count events and also time them.[7] The fact that the same mechanism could do both is necessary for animals to estimate rates (time/numerosity) – that is, counting and timing are measured in a 'common currency', the level of the accumulator. Of course, one obvious way to calculate rates is if both time and numerosity are represented by numbers. Their paper was called 'A mode control model of counting and timing processes' and argued that the accumulator mechanism's 'mode' of operation could be switched between counting and timing.

In these experiments, the rats listened to a sequence of sounds, of variable duration and numerosity. They could distinguish a 4:1 ratio of counts (with duration controlled) and a 4:1 ratio of durations (with number controlled) – see Figure 1.

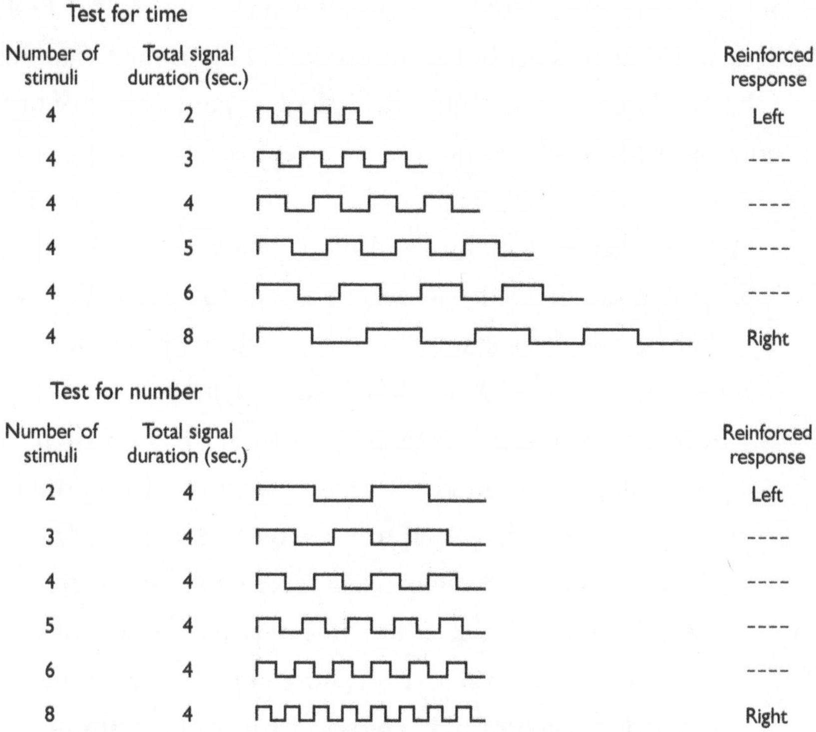

Test for time

Number of stimuli	Total signal duration (sec.)		Reinforced response
4	2		Left
4	3		----
4	4		----
4	5		----
4	6		----
4	8		Right

Test for number

Number of stimuli	Total signal duration (sec.)		Reinforced response
2	4		Left
3	4		----
4	4		----
5	4		----
6	4		----
8	4		Right

Figure 1. The number of bursts of white noise is indicated by the raised humps. In the 'Test for Time' (duration), the number of bursts is constant at four, but the duration varies from two to eight seconds. In the 'Test for Number', the duration is held constant but the number of bursts varies from two to eight.[7]

The rat had a choice of right lever or left lever. In the test for number, the right was rewarded for eight sounds, and the left lever for two sounds, with the total duration of the stimuli constant at four seconds. In the test for time, there were always four separate sounds but the total duration varied, and with the left lever press rewarded for a two-second duration and the right for an eight-second duration. The results were clear. The rats could learn to discriminate both duration and number. The discriminability of both dimensions was almost identical in the two tasks, suggesting that the same mechanism, the accumulator, was used for both. Meck and Church describe the mechanism of timing and counting in the discussion of the results. This is the first presentation of their enormously influential accumulator model (see Chapter 1):

> A pacemaker puts out pulses. A mode switch can be closed to pass these pulses to an accumulator. The pacemaker-switch-accumulator system may be called either a clock or a counter. It is used as a clock if the switch operates in a run or a stop mode; it is used as a counter if the switch operates in an event mode. In either case, the value in the accumulator may be passed to working memory . . . The current accumulator value is compared to the remembered accumulator value at the time of reinforcement of a previous response, a value that is stored in reference memory. The decision process is a response rule that determines the response.

Of course, it was possible that in the rat's brain there were two very similar mechanisms that behaved in identical ways, one for duration and the other for counting.

Since this ground-breaking study, there have been many follow-ups and replications, and not only in rats. In the next chapter we'll see that birds do at least as well on these counting tasks.

More recently, interest has focused on the numerical abilities of mice. Scientists have been studying lab mice for more than a hundred years, so we know a great deal about their behaviour. Since they don't live long – a couple of years in the lab, less in the wild – they are good candidates for developmental studies. One other, increasingly important reason for choosing mice is genomic.

Almost all the genes in mice share functions with the genes in humans, and there are now many strains (genetic variants) that allow scientists to test the role of these genes in disease and in behaviour.

There are few relevant studies of mice in the wild, but a version of spontaneous wild behaviour was reported by Sofia Panteleeva and her colleagues from Novosibirsk in Russia. Striped field mice (*Apodemus agrarius*) hunt and eat ants, but they know that ants can sting, and that lots of ants can sting painfully. In this study, wild and lab-reared mice were placed in an arena and could see ants in two transparent tunnels such that the number of ants differed – 5 vs 15, 5 vs 30, and 10 vs 30 ants. It turned out that the mice, both lab-reared and wild, almost always preferred the smaller number. The title of the paper sums it up: mice 'first "count" and then hunt'.[8]

I want to describe a beautiful recent study by two Turkish scientists, Bilgehan Çavdaroğlu and Fuat Balcı, that uses Mechner's paradigm but shows something new and very important. They trained each mouse on three fixed-ratio schedules, which they call 'fixed consecutive number' (FCN) schedules, so that they had to press a lever ten, twenty and forty times, before switching levers to obtain a reward. Can mice with their relatively small brains manage to count that high? It turns out that they can, with very similar responses to Mechner's original study with rats.[9]

The thing to note is that spread of responses – variability – increases with the number of lever presses required. That is, there is scalar variability, which in this study was the same for the three schedules. Now it's not that the mouse simply has only an approximate sense of the number of lever presses needed. It turns out that the peak of the distribution is greater than correct response for each target numerosity. And it is proportionally greater for then 40 schedule than the 20 schedule, and proportionally greater for the 20 schedule than for the 10 schedule. Çavdaroğlu and Balcı argue that this is because the mouse somehow has access to its internal uncertainty and takes this into account when pressing the lever. To be sure of getting a reward, it is better to press too many times than too few, when you are not rewarded, so when the estimate of the internal uncertainty is large it is better to err by making more lever presses. However, there is a cost to the extra effort to these superfluous lever presses. Çavdaroğlu and Balcı have calculated the optimum balance of reward and cost, given the mouse's estimate of the uncertainty, and it turns out that the mice are behaving in close to the optimum way.

Whales and other cetaceans

These excellent mouse counters have brains that weigh less than half a gram. Whales, by contrast, have massive brains. Sperm whales (*Physeter macrocephalus*), for example, have the largest brains of any animal on earth, weighing in at 7.8 kilos in mature males. Humans have brains of about 1.4 kilos. Some also have, of course, the most massive bodies of any creatures to have walked or swum the earth, and the greater proportion of brain mass to body mass may mean the greater amount of brain mass that may be available for cognitive tasks, such as counting. The other issue is the location of the neurons in the structure of the brain.

It is generally believed that the neurons that do the heavy cognitive lifting in human brains are in the neocortex, the thin layers that cover the surface of the brain, and that the parietal lobe surface neurons in humans are the hub of our numerical abilities.

Now it is difficult and probably very expensive to get hold of whale brains and also difficult to count the number of neurons in different brain regions. One team based in the Faroe Islands, a big whaling centre, counted the neocortical neurons in the long-finned pilot whale (*Globicephala melas*), which have brains weighing between 3 and 4.6 kilos.[10] They estimated the number of neurons in the neocortices of ten juveniles and adults at 37 billion, nearly twice as many as humans! Also, their brains are more folded than ours, which means more cortical surface area, which in turn is thought to reflect cognitive abilities: so the more folds, the better. 'The area of the human neocortical surface is 2,275 cm² (about the size of a dinner napkin), but the common dolphin neocortical area is 3,745 cm² (bigger than an unfolded newspaper).'[11] The cetacean brain also contains spindle cells (von Economo cells), which are neurons essentially with one input and one output branch that are found almost exclusively in humans and great apes, and may play a key role in social intelligence.

Does this mean that these cetaceans are at least as smart as us? Probably. Cetaceans show sophisticated social behaviour, making and perhaps breaking alliances; they cooperate when hunting and share hunting techniques; they produce complex vocalizations, with regional dialects; they can share parenting duties and enjoy social play,[12] they even collaborate with other species, including humans, in foraging,[13] and all rather peacefully.

Perhaps unsurprisingly, the little work that has been done on the numerical abilities of whales and other cetaceans such as dolphins has been carried out in aquaria.

Nevertheless, there is indirect evidence of numerical abilities in the wild. Humpback whales (*Megaptera novaeangliae*) migrate up to 5000 kilometres each year after feeding up on krill in the cold waters of Antarctica to prepare for the long journey ahead and stopping to breed in the warm waters of the Great Barrier Reef.[14] In fact they repeat the same route each year with extraordinary fidelity. To manage this feat of navigation, they calculate the route using the magnetic field of the earth, the position of the sun, moon and stars, the undersea geography such as sea mounts where they stop to snack, temperature gradients in the ocean and maybe other clues to where they are and where they need to go next. This is an amazing achievement and probably involves a great deal of calculation of direction and distance, what sailors call 'dead reckoning' and scientists call 'path integration'. Picture, if you will, the navigator on board a vessel with a map, compass, protractor, ruler and chronometer to plot the route. The whale does all this in its massive head without benefit of these tools. One of the main brain structures for navigation in mammals and birds is the hippocampus, and cetaceans have massive hippocampi, some larger than humans.[15] (I deal with navigation in more detail in Chapter 9 on insect navigation where the mechanisms are better understood.)

There have been a few lab-type experiments with cetaceans housed in aquaria. One very nice example set a beluga whale (*Delphinapterus leucas*) and three bottlenose dolphins (*Tursiops truncates*) a simple task to choose one of two boxes with more fish.[16] They were rewarded with the fish in the chosen box. Comparisons of one to six fish were tested, and all did quite well, though not completely in line with Weber's Law. So the bottlenose dolphins happened to do better on 4 vs 6 than on 1 vs 5. In one fascinating condition, the beluga whale couldn't see the contents of the boxes, but had to estimate the contents using a kind of sonar. Since they live much of the time in pitch black, they use a system not unlike that of bats, emitting a rapid sequence of clicks, and they can interpret the echoes of the clicks as, among other things, food objects. However, it is worth noting that whichever box the whale chose, it would be rewarded.

A more stringent test was carried out with a six-year-old bottle-nose dolphin called 'Noah' at the Nuremberg Zoo by a German team led by Annette Kilian.[17] After training to select the smaller of two sets of three-dimensional objects, Noah had to select the smaller set of two-dimensional objects (Figure 2).

In this way, it was possible to control other visual variables so that success could be confidently ascribed to numerosity and not

Figure 2. The basic set-up for the dolphin task to select the array with the larger number of elements.[17]

other visual features, such as the overall surface area of the stimuli and using heterogeneous objects of different sizes. Noah was well able to select on the basis of numerosity with these controls in place. Noah could also transfer to pairs of numerosities not previously presented, even 5 vs 6. Maybe we shouldn't be surprised that dolphins with their big brains – at least as big as humans and a better brain-mass-to-body-mass ratio – should be able to make these numerical discriminations, but it is worth remembering that they live their life in a very different – more fluid – environment than land animals.

We are still waiting to discover the upper limit of cetaceans' counting ability, and their ability to carry out calculations.

Studies with captive mammals

Lab studies show that many mammalian species can make discriminations based on numerosity. Dogs will choose the larger number of food items, with Weber Law effects; similar effects have been observed in wolves and domestic cats, and in captive sea lions, elephants, coyotes and black bears.[4] The last is of particular interest, at least to me. I was once asked by a student whether all creatures with numerical abilities were social since all the examples I gave were from social animals and maybe that was why numerical abilities were important to them. This was a very good question and at the time, I had no answer, as I rather sheepishly admitted. Now I know the answer. Sarah Benson-Amram and her colleagues write, 'The American black bear, a solitary carnivore, is capable of discriminating these quantities [of dots], even when presented with moving stimuli. Thus, this ability did not only evolve in social species and is not solely an adaptation stemming from the need to keep track of group members.'[4]

From these animals with very big brains, back to rats. Rats can easily learn to select a target box containing food based on its position in an array of several identical boxes (up to the twelfth position is a sequence of eighteen boxes). That is, they count up to twelve one by one, and use the remembered information to find the food. In terms of the accumulator mechanism, they have to have a way of storing the accumulator level with great fidelity for later use.[18] Extraordinarily, rats can remember this information for more than a year.[19]

Cat owners look away now. Way back in 1970, Richard Thompson and his colleagues at the University of California Irvine investigated the neural basis of numerical abilities of the cat.[20] To do this he implanted electrodes in the brains of anaesthetized cats, into what was called an 'association area' because these are areas of the cortex that respond to stimuli in different modalities and can link them together. Happily, it turned out that the region of the association cortex was the parietal lobe. He recorded the response of individual neurons to the presentation of a sequence of clicks or light flashes on the assumption that both modalities would be represented in this association area, and that maybe exactly the same neurons would respond in the same way to the same numerosity, whether the cat heard or saw the set of stimuli.

The clicks or lights could be presented at a one-second interval in the auditory or visual sequence, or at an interval of four seconds in the auditory sequences. Briefly, they found that cells coded for particular numerosities up to seven, *irrespective* of modality or interstimulus interval (see Figure 3, overleaf).

Thompson and colleagues observed five 'counting' cells which code for two, five, six and seven stimuli, and they conclude that

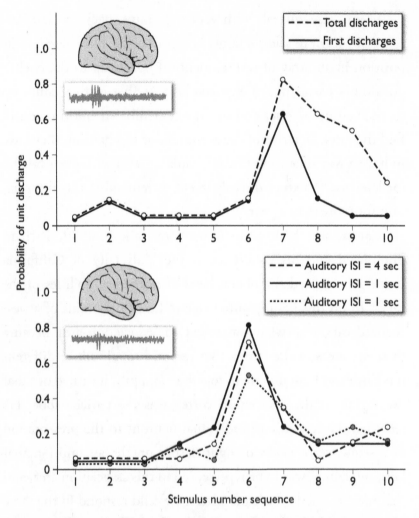

Figure 3. The probability of a response in an individual neuron in the cat's parietal lobe. In the top panel, this neuron is much more likely to respond for a sequence of seven stimuli than for any other numerosity. The lower panel shows the probability of response for an individual neuron for a presentation of six stimuli, irrespective of the modality of presentation, auditory or visual, or the interval between clicks in the auditory sequences.[20]

'The "counting" cells described here behave as though they code the abstract property of number.' These cells were triggered by discrete events and fired only when their respective target number was reached.

In terms of our proposed accumulator mechanism, the selector would open the gate to the accumulator for either auditory or visual events, but the 'seven neuron' would fire only when the accumulator reached a pre-defined level.

We have seen that in the wild, lions and hyenas use abstract numerical assessment to minimize lethal inter-group conflict. In lab-like conditions, rats can count sounds in order to get a reward, and mice can count up their own lever presses to at least forty in a way that optimizes the cost–benefit ratio of over-counting. These studies, and the responses of cat parietal neurons, suggest that all these creatures are born with an accumulator kind of mechanism in their brains that can respond to numerosities in different modalities and modes of presentation. It is possible, even likely, that this mechanism has its origin in the common ancestor of all mammals.

CHAPTER 6

CAN BIRDS COUNT?

Birds, with tiny brains that have to fit into their small, aerodynamically shaped heads, turn out to be champions of the animal number world, as we will see. As well as identifying numerosities accurately, they can carry out the same kind of calculation as young human children. More remarkably, perhaps, they can navigate thousands of kilometres over bare ocean, and find their way back home. We now know something about how they do it, but not – at least not yet – the computations that are required to plot these journeys.

Learning to count

There is one truly amazing case of bird counting, and that is Alex the Parrot, trained for thirty years by Irene Pepperberg, initially at the University of Arizona, later at Brandeis University and Harvard. Alex's death in 2007 was worthy of extended coverage in the world's press and media, including *The Economist*, the scientific journal *Nature*, and the *New York Times*, which headlined its obituary 'Alex, a Parrot Who Had a Way With Words, Dies'. Its author, Benedict Carey, described how Alex was trained:

African Greys are social birds, and pick up some group dynamics very quickly. In experiments, Dr Pepperberg would employ one trainer to, in effect, compete with Alex for a small reward, like a grape. Alex learned to ask for the grape by observing what the trainer was doing to get it; the researchers then worked with the bird to help shape the pronunciation of the words. (10 September 2007)

This is called the 'model/rival technique' and is often used to train other animals after Pepperberg's success with Alex.

Alex also had a very distinctive personality. If he became fed up after repeating learning trials dozens of times, he would throw objects off the stimulus trays with his beak, and demand to go back to his cage. When Alex died in September 2007, his last words to Pepperberg were 'You be good. I love you.'

What is remarkable is that Alex was able to do things that were thought to be uniquely human; most particularly, as the *New York Times* noted, he could speak and understand many words. After extensive training, Alex could identify fifty different objects, seven colours and shapes, and could also name them. His vocal and communicative abilities were also remarkable. He seemed to understand what was asked of him. He responded correctly when asked about the shape, colour or material of an object on a tray. He would call a key a key no matter what its size or color.

There's no doubt that Alex was smarter than the average grey parrot (*Psittacus rithacus*). He did much better on the vocabulary tasks than other parrots in the lab. For example, a twelve-year-old parrot named Griffin mastered fully only twenty words. Mind you, Alex sometimes practised words when he was alone.

His cognitive development included 'object permanence' – that is, he had expectations that objects would continue to exist even when they became hidden from view, so that he showed surprise and even anger when confronted with an unexpected object in the reveal. The influential Swiss psychologist Jean Piaget found that human babies only reach this stage of cognitive development by about eight months.[1]

Alex could learn the abstract relational concepts of *same* and *different*. When he was asked 'What's same?' or 'What's different?', he 'replied with the correct English categorical label ("color", "shape", or "mah-mah" [material]) about pairs of objects on the tray that varied with respect to any combination of attributes. His accuracy was 69.7%–76.6% for pairs of familiar objects not used in training and 82.3%–85% for pairs involving objects whose combinations of colors, shapes, and materials were unfamiliar.'[2]

But it is his numerical abilities that are of special interest to us. He could name quantities of up to at least six. He learned the relationship between the spoken number words and sets of objects, and later the relationship between the digits and sets.

In one early study, Alex learned to name ('label vocally') collections of one to six objects of the same type on a familiar tray, and then, without further training, the number of subsets. So, he was presented with different collections of '4 groups of items that varied in 2 colors and 2 object categories (e.g., blue and red keys and trucks) and was asked to label the number of items uniquely defined by the conjunction of 1 color and 1 object category (e.g., "How many blue keys?").' On this task he was remarkably accurate, at least 66 per cent correct, and on the four-item condition, he got nine out of nine correct.[3] This is important because it demanded quite sophisticated use of the selector to deliberately pick out just the blue keys and increment the accumulator appropriately. This also implies a degree of numerical abstraction. The word 'six' – either spoken by Alex or heard by him – applies to any set of requested objects on the experimental tray.

As I specified in Chapter 1, to have the ability to count, you have to be able to do something with the result of counting that is 'isomorphic' with arithmetical operations. Of course, many of the studies I have described in earlier chapters involve the operation of comparing the numerosities of two sets A and B (A < B, A > B, A=B). In some cases, simple calculation (A + B, or A – B) is involved in lab studies. It is worth reminding ourselves that even numerosity comparison can involve adding. When a set is observed, its numerosity may have been established by sequentially adding up the objects, $1 + 1 + 1 \ldots$, and in some experiments, the animal is presented with the objects one at a time, so adding is entailed in making the comparison. We have seen that primates can carry out these operations even without training. What about Alex?

Alex could indeed add sets. To understand what was involved in the addition task, I need to go into some detail. The objects to be added were familiar treats (such as nuts). Two sets of nuts were placed on the familiar tray and each set covered with a plastic cup, A and B. Alex was shown what was under cup A, which was then replaced, and then under cup B, which was then replaced. This meant he could not see the nut pieces. He was then asked 'How many nuts total?' Pepperberg points out that 'To respond correctly, he had to remember the quantity under each cup, [and] perform some combinatorial process.' There was no training.

Various other familiar treat objects were also used in sums adding up to between 1 and 6. On all the sums, Alex scored 7/8 or 8/8, apart from 5 (4/8). When he did make an error, he didn't just pick a random number, but labelled one of the sets under a cup, suggesting he hadn't carried out the addition but was focused on one of the sets.[4]

Could Alex understand digits, just like children at the end of Grade 1? Pepperberg and Harvard developmental psychologist, Susan Carey, investigated this. We have seen that he could vocally label sets of 1 to 6 objects. He was then taught to label the digits 1 to 6. In this task, the digits 7 and 8 were also used.[5] The model/ rival method was again used to train Alex the orderings: '6' < '7', '7' < '8', '6' < '8', '8' > '7', '8' > '6', '7' > '6'. The digits were given a variety of colours on the tray, modelling replies to 'What colour number (is) bigger/smaller?' to teach ordinal relations among '6' ('sih-sih', Alex's 'six'), '7' ('sih-none', Alex's 'seven'), and '8' ('eight'), to provide evidence for the numeral status of '7' and '8'.

Alex was highly accurate on these ordinal tasks, even those with the new symbols 7 and 8. The experimental digits came in different colours, and Alex correctly answered 'What colour 7?' when the digit was orange, yellow or blue.

One other result of these tasks was very interesting. After Alex had labelled a set, one or two objects were added or taken away from the set. All but one time, he gave the correct number to the transformed set. For example, when presented with a set of seven blocks, he said 'seven' and when one was added, he said 'eight'. And when two were removed from a set of eight blocks, he said 'sih-sih' (six).

He seemed to understand that transformations that change the numerosity – adding and subtracting – require a change of verbal number label. This is a bit like an experiment with three- and four-year-old human children I described in Chapter 2,[6] except here Alex didn't just give a different number as many of the children do, he gave the correct number. He seems to have had what Piaget would call conservation of number, though Piaget would expect this to be realized in children nearer to seven years old.

Pepperberg and Carey concluded that Alex understood 'the successor function', in the sense of an ordered series of numerals and next in a series of sets, n and $n + 1$ and so on, rather than just having a sense of approximate n-ness. 'These are among the logical foundations of integer concepts; Alex's deployment of them undermines claims of human uniqueness with respect to these computational resources . . . Two considerations lead us to favor the hypothesis that some representations based on 1–1 correspondence and "exactly N" (rather than "approximately N") supported the meanings Alex assigned to numerals.'

Just before Alex's death, Pepperberg started to teach him to add digits. Could he possibly do it? As in the addition task with objects, two digits were presented and then covered with cups. The digits were in the range one to five, and the sums up to eight. Digits in different colours were visible on the experimental tray, and he was asked to name the colour of the sum of the two digits concealed by the cups. Out of fifteen trials, which was all that could be carried out, he got twelve correct.[7]

It turns out that parrots and corvids (crows, ravens, jackdaws) are better at using numbers than any other species, apart from chimps. The best early research on animal number abilities was in fact on birds. Otto Koehler devised methods that set the standard for properly controlled experiments of the numerical abilities of a variety of birds and some mammals, as I mentioned in Chapter 1. Nevertheless, he believed that although animals could learn to use numerosity in the lab, they were not able to use numbers in the wild.

Koehler's preferred method was 'match-to-sample'. That is, he would present the bird with a sample of a particular numerosity, and his corvids were able to match numerosities up to seven. For example, with jackdaws he would present a sample number of ink dots (or other objects), and the bird's task was to find the lid of a box with the same number of dots. If the bird found the correct box, lifting the lid would offer a food reward.

As I pointed out in Chapter 1, it was vital to ensure first that the bird was using numerical information and not other visual information, such as the total area of ink or total quantity of object stuff. Koehler did this very systematically by ensuring that the tasks could not be solved on the basis of total area etc. Figure 1 shows examples of this method.

He also used an interesting variant, making the sample sequential. A jackdaw had to open boxes one by one until it had obtained five baits – the sample – which were distributed one in the first box, two in the second, one in the third, none in the fourth, and one in the fifth, of eight boxes in all. Then the bird should return to its cage to indicate that it has completed the task. In one unusual and

Figure 1. Some of the stimuli used by Otto Koehler when testing birds, the basic set-up for a match-to-sample study. The bird has to match the big stimulus to one of the two smaller matches.[8]

extraordinary trial, it went back to its cage. Koehler was about to record 'incorrect solution, one too few', when the bird came back to the line of boxes. The jackdaw went through a

> most remarkable performance: it bowed its head once before the first box, made two bows in front of the second box, one before the third, and then went farther along the line, opened the fourth lid (no bait) and the fifth and took out the last (fifth) bait. Having done this, it left the rest of the line of boxes untouched and went home with an air of finality.[9]

In another paradigm Koehler devised, a bird had to collect baits from n boxes and then find a box with n dots or bits of plasticine on the lid to receive its reward. This implies that the bird is carrying out sequential counting, just as we would do, and retains the total number of objects counted in memory to do the next part of the task. This means that sequential and simultaneous sets were represented in the same way in the brain of the bird. That's pretty abstract.

Although I have argued that Koehler's birds were really counting, he himself believed that his birds were not really counting, but used 'unnamed numbers' in their brains, where the numbers consist of *equal marks*. Notice that if each object was represented by *unequal* marks then the enumeration would depend on the nature of what was being enumerated. The equal marks are very much like the 'numerons', the innate internal representations of numerosity, proposed by Gelman and Gallistel in their ground-breaking book about human children, *The Child's Understanding of Number* (Chapter 2). That is, with equal marks, whatever the animal had to enumerate, each object was internally represented in the same way. This is true for the 'marks' of objects presented both simultaneously and sequentially. This is exactly the way the accumulator works. Each object or event elicits the same unit of activation. The jackdaw bowing before the boxes was, according to Koehler, *thinking unnamed numbers* in order to determine whether the sample had been correctly matched. Nevertheless, Koehler was not convinced that the numerical capabilities revealed through training could be observed in the wild – a point to which I will return, since it is clearly important to the evolutionary claim that our numerical abilities are based on an inherited mechanism.

Koehler worked with various birds including grey parrots, budgerigars, crows, ravens and magpies, and even squirrels. He was careful to avoid Clever Hans-type problems (see p. 24) by ensuring that he and other experimenters stayed out of sight and that the birds were filmed to provide an objective record of their behaviour.

Following Koehler's ground-breaking studies, new experimental paradigms were devised. One neat example comes from a study of 'Jako', another grey parrot. He was able to match a sequence of light flashes to a sequence of actions, taking baits out of a row of food trays. Not only that, but when the light flashes were replaced by a sequence of notes of a flute, Jako was able to transfer without training to match the number of notes with the number of actions.[10] This is a nice example of numerical abstraction – equivalence across modalities, sounds and actions.

More recently, Andreas Nieder, who we saw in Chapter 4 training monkeys to match numerosities, and who discovered 'number neurons' in their brain, has turned his attention to these very clever corvids, in this case crows (*Corvus corone corone*), one male and one female. He used an almost identical experimental design to the one he used with macaques (see Chapter 4): match-to-sample with many controls to ensure that the subject really was using numerosity rather than any of the other visual dimensions that typically co-occur with numerosity: more dots = more area unless dot size is controlled, and so on (see Chapters 1 and 4). Using modern technology to update Koehler's methods, the crows had to peck a touch screen to make the match (see Figure 2).

It turns out that the crows are pretty good at this, and always above chance at matching one, two, three, four or five dots. This was the case whatever the controls, showing that these birds really can use numerosity to do the task. In fact, the accuracy of the crows was almost identical to that of the monkeys.

Figure 2. Match-to-sample with a crow.

Crows are not the only birds on a par with primates when it comes to numerical ability. In a landmark study, Elizabeth Brannon and Herb Terrace showed that monkeys could be trained to order sets of variegated stimuli from one to four in ascending order. They ensured that the monkeys were using numerosity, not any other visual feature, by controlling or randomizing the size, shape and colour of the objects in the set. They then tested the monkeys on pairs of novel numerosities outside the training range, which they could easily manage and which followed Weber's Law – that is, accuracy depended on the proportional difference between the pairs of sets (see Chapter 4 for a description of the study, and examples of the stimuli).

Pigeons can do the same as the monkeys. Damian Scarf, Harlene Hayne and Michael Colombo from the University of Otago in New Zealand borrowed the stimuli Brannon and Terrace used with monkeys in a study of three pigeons and found almost identical results.[12]

Brannon commented, 'I thought it was amazing that monkeys could do this, so we should be even more impressed that pigeons can, too!'

These tests involved a lot of training, but there are studies in which no training at all was required to demonstrate numerical abilities. Such research is particularly important because it establishes what comes naturally and what might be observed in the wild.

Doing what comes naturally

Many birds are known to 'imprint' on the first thing they see on hatching. They follow it, since the first thing they see is likely to be a nest mate or its mother, and so this will help its survival. Imprinting in geese is associated with the 1973 Nobel laureate Konrad Lorenz (1903–1989), but in fact has been known since ancient times. Pliny (23–79 AD) speaks of 'a goose which followed Lycades as faithfully as a dog',[10] and several early naturalists, not to mention Sir Thomas More in *Utopia* (1516), described the phenomenon. Lorenz, however, observed that imprinting could occur when the newly hatched gosling has only a few minutes to notice something moving.

Now suppose a newly hatched bird sees two nest mates; will it then tend to follow two objects rather than one or three? Or, without nest mates, if it sees one object moving left and two objects moving right, which will it follow? The key numerical question is whether it can distinguish a set of one from a set of two independently of the members of the set. Suppose the bird sees four going right and one going left, and then three of the right-movers then go left, making a set of four going left and only one going right, will it be able to calculate that there are now more going left?

These are questions that Giorgio Vallortigara, Lucia Regolin and their colleagues asked of newly hatched domestic chicks (*Gallus gallus*). Rosa Rugani, originally a student of Regolin, gave a brilliant and crystal-clear presentation of the team's work at the famous Royal Society meeting on the origins of numerical abilities in 2017, which Giorgio, Randy Gallistel and I organized.[13]

Rugani described a study in which chicks were reared with different numbers of objects that varied in shape, size and colour. One group of chicks was reared with a single object and another group with three identical objects.

At test, completely novel objects were used, different from the rearing ones in colour, shape and size. Chicks approached the numerosity they were raised with, even though the objects are quite different, suggesting that whenever no other cues are available, chicks can rely on number, and this must be a capability they are born with as there was no training, and no opportunity to learn.

It turns out that when moving sets of plastic balls are visible, chicks prefer the larger set. A later study tested chicks' ability to remember the number of plastic balls after they had disappeared behind a screen and, given a free choice, approach the larger set (3 vs 2).[14]

Even more remarkable than an innate ability to discriminate remembered numerosities is their innate ability to carry out calculations. There were two conditions. In the first, four balls disappeared behind one screen, and one behind the other. The two balls were moved from the first screen to the second in full sight of the chick, which was then let out of the holding box and could approach either screen. The chick could not see what was behind the screens and had to rely on her memory of the events. Significantly often, she approached the screen which now had three balls behind the screen. This implies that she had to subtract two from four and add two to one. In a second condition, there were five balls moved behind one screen and none behind the other; then three were moved from the first to the second. To approach the screen with more balls, the chick had to subtract three from five and add three to none. To ensure that the chicks were calculating rather than just approaching the last ball that moved, a control condition was employed in which following the last ball would not be the screen behind which were more balls. In both conditions, the chicks were approaching the screen with the most balls 70 per cent of the time.[14]

Notice that the chick starts in a transparent holding box so she can see objects moving behind the screens on the right, and also see the objects moving from one screen to the other. When she is let out of the box, all the objects are concealed behind the screens, so a decision must be made on the basis of her memory of events.

Using numerical abilities

The lab studies suggest that birds can learn to use an innate to ability count. But why should birds be born able to count? What use is it in the wild? Followers of Darwin, such as William Lauder Lindsay (1829–1880; see p. 310), argued that 'lower animals' had a 'faculty of numeration'. George Romanes (1848–1894), who had worked with Darwin, reported an anecdote from a ranger at the palace of Versailles whose job included shooting pests, which included crows. The ranger discovered that the crow would not return to its nest if it saw a hunter entering a blind. In order to fool the crow, six men would enter the hide, and then five would leave and the hunter remain because, they thought, this would 'put her out of her calculation'.[15]

Counting ability turns out to be advantageous in different ways for different species. For example, American coots count their own eggs to calculate how much food to forage.[16] Nest parasitic birds (such as cowbirds) use the number of eggs already in the host's nest to time their own egg-laying: the female cowbird will lay close to the typical maximum number for the host species to begin incubating.[17] Songbirds also need to be able to count notes in their own songs and the songs of competitors. Many songbirds learn their song by imitating those of older members of their own species, probably by modifying a simple innate song until auditory feedback matches a heard and memorized model. This ability seems to be a specific cognitive capacity, not just general bird intelligence,[18] and is implemented via a specialized neural network in the pallium (about which more below).[19]

Songbirds, for example the swamp sparrow (*Melospiza georgiana*), also notice when an extra note is added to a song. This is important because it means that the song is not from neighbouring males and could constitute a threat.

Finding the way home

Birds forage at considerable distances from their nests, and some, of course, migrate over hundreds or even thousands of miles. This extraordinary behaviour includes complex computations involving what sailors would call 'dead reckoning', which for humans typically involves a map, a compass and a calculator, so that the distance of each turn can be plotted in order to compute where the boat is on the map. In the case of birds, they also have to return to their nest after foraging. So not only do they have to figure out where they are, they also have to determine the 'home vector', the direct route back to the nest. Two key elements are a compass and a map. The bird's compass is known to depend on many different types of information: the position of the sun, the polarized light pattern in the sky, especially when the sun isn't visible, the stars at night, maybe wind direction. The map will encode landmarks, and sometimes smell beaconing to the home burrow for seabirds once they are very near the nest site; many birds use a mechanism in the eye for magnetic reception as a way of estimating latitude, which is important for long-distance migrants.

The other element in route finding is distance travelled. Sometimes familiar landmarks help provide this information. This could be estimated by what are called 'idiothetic cues' – information from the bird itself. These could include optic flow to estimate speed of travel, or energy consumption as a way of estimating distance. How these sources of information are deployed will depend on whether the bird is foraging near the nest, or returning from a longer excursion, like homing pigeons, or migrating over very great distances.

I have to say that I find these long-distance migrations staggering. The bar-tailed godwit (*Limosa lapponica baueri*) breeds in Alaska and winters in New Zealand. That's an annual round trip of 29,000 kilometres! Non-stop from Alaska to New Zealand in eleven days – and that's with a 5 gram satellite tag on its lower back to allow scientists to track its progress.[20]

Not only is the distance enormous, but it is across the Pacific Ocean, which the great godwit experts describe as 'the most complex and seasonally structured atmospheric setting on Earth'. To make the journey, godwits need to maintain a bearing day and night, which implies that they have to use a solar compass during the day and sky compass at night, adjusting for the ephemerides (changing position of the sun and stars over time) of the sun and stars. It is thought they also have a magnetic sense which modelling indicates contributes to estimating the bearing.[21] Plus they have constantly to take into account wind speed and direction. And that's just in two dimensions. They also have to compute the best height for their flyway. And there are many birds which make these long migrations, and the arctic tern (*Sterna paradisaea*) that lives in Greenland and Iceland, makes an even longer annual migration to the Antarctic of 70,000 kilometres on wings of no more than 40 centimetres, returning to the same colony each year.

According to the great bird navigation expert, Dora Biro of Oxford University, 'birds predominantly rely on map-and-compass navigation'. They have a map in their heads that allows them to fix their current position in space and also their position relative to a target (and they can then use their compass to take up the direction they calculated between current position and the target).

Biro tests this by carrying out:

experiments with homing pigeons (*Columba livia*) where we 'clock-shift' them (i.e. they're jetlagged, which causes them to misinterpret the sun's position, and that in turn gives them an erroneous, shifted geographical compass direction). When clock-shifted pigeons are released, they will initially fly off in the wrong direction (thinking that they're heading home) but there is some anecdotal evidence that they begin to realize something is wrong once they've flown a distance that is roughly equivalent to the distance it would have taken them to get home. So that would suggest that they had some expectation about the distance that they were from home at the outset (despite having been transported to the release site in a car, rather than navigated there by themselves). This could work if, for example, they have some sort of bicoordinate map, on which they have 'home' marked with coordinates, and then they can also work out the coordinates at the release site. Or it could have been learnt through experience of previously having homed from the same site and remembering the distance flown. (personal communication)[22]

Manx shearwaters (*Puffinus puffinus*) seem to know both the direction and the distance to home when they set off back towards the colony since they leave earlier in the day when they have further to go.[23]

A beautiful study with white-crowned sparrows (*Zonotrichia leucophrys gambelii*) showed how maps and map directions are learned. The sparrows in question migrate in large flocks of experienced adults and naive juveniles from Washington State to Southern California and Mexico. In this study, the birds were transported in mid-September by plane to Newark, New Jersey, in a compartment controlled for temperature and pressure with no windows but continuous dim lights.[24] They were then transported by car to Princeton, New Jersey, where they were housed in groups of three in laboratory bird cages until released.

Now when they fly from Washington State they head more or less due south. What will they do when released 3700 kilometres east of this? It turned out that adults who had flown the route before, actually headed south-west to the wintering grounds while naive juveniles flew due south. This showed that 'the learned navigational map used by adult long-distance migratory songbirds extends at least on a continental scale. The juveniles with less experience rely on their innate program to find their distant wintering areas and continue to migrate in the innate direction without correcting for displacement.'[24]

These journeys require complex and sophisticated calculation – think James Cook's voyages to the Pacific – and exactly how they do it in real time remains to be determined. I asked a sailor friend, Kevin Brown of Molokai, Hawaii, how distance would be estimated without modern equipment. Maybe this would give us a clue as to how birds do it. He replied, 'Distance travelled can be calculated using a *chip log*, i.e., timing the passage of an object alongside the moving vessel, but this gives only the vessel's speed and so distance travelled is, again, a matter of calculation.'

Michael Hammond, a historian of science, filled in some of the details:

> A chip log was a simple device that consisted of a piece of wood attached to a rope that was wound on a reel. The rope had a knot tied in it every 47.3 feet. To use the log the wood would be thrown overboard and a sand-filled timer, much like an hourglass, would be turned over. The timer counted down 28 seconds, we wanted to leave the log out for 30 so we took off two seconds to account for the time it took us to throw the log and flip the timer. The number of knots that passed through our hands was equal to the number of nautical miles per hour that we were traveling, and there is the way that the term knot came into our vocabulary. The reason that this works is simple math. Instead of leaving the rope out for an hour and counting how many miles of rope were let out we simply scaled it back to a practical amount of time and distance. There are 120 thirty-second intervals in an hour and 120 47.3-foot intervals would equal about 1 nautical mile. Although it doesn't work out to be exact it is pretty close and certainly close enough for us sailors.[25]

We know birds can estimate speed using optic flow – the rate at which objects, or the sea, moves across the eye's retina. One of the clocks in the bird's brain could be used to estimate distance as a function of speed and time. Like the sailor's chip log, these estimates are necessarily approximate, and it is likely that estimates of distance will be checked against other information, including, where possible, landmarks, or celestial clues.

One intriguing possibility of how birds and other creatures, including insects as we shall see, compute path integration has been suggested by Thomas Collett at the University of Sussex. Imagine not one accumulator, but:

> a spatial array of accumulators with different preferred heading directions, each updating according to the insect's current direction of travel. The first of these models . . . considered the simplest case of two accumulators, one integrating components of travel in, say, the east–west direction, the other in the north–south direction. During any outward path, the two accumulators sum their independent components. The integrator has no history of how the end state was reached, but the vector sum of the contents of the two accumulators generates a home vector.[26]

One way of thinking about navigation in terms of numbers has been proposed by Gallistel. Just three 'bits' of information – 000, 001, 010 etc – can represent the magnitudes of turns to an accuracy of one-eighth of a turn, that is, the eight principal points of the compass ('N', 'NE', etc.).[27]

Gallistel explicitly compares sailors navigating and animals navigating. This will crop up again in Chapter 9 on bee navigation.[28]

> Latitude and longitude are not interchangeable in navigational computations, nor are range and bearing. The four most common navigational computations defined

on spatial vectors are vector subtraction (in course plan-
ning), Cartesian to polar conversion (in course setting),
polar to Cartesian conversion (computing the location to
which a given course, i.e., rhumb* line, will take you),
and vector addition (in dead reckoning, and more generally, in determining to what location a sequence of courses
takes you).

This is what is needed for map-and-compass navigation, and
it entails that the navigator, bird or human, has a map. In the case
of the sailor, this could be a paper map; in the case of the bird, or
bee, or the early Polynesian explorers of the Pacific, the map will
be in the brain.

Actually a normal type of map may be necessary, but is not
sufficient because it is only in two dimensions. Birds fly in three-
dimensional space to evade a predator or to catch a prey, find the
best altitude for migration or flying to nest in a tree or a cliff. The
bar-headed goose (*Anser indicus*) migrates from its summer habi-
tats in Kazakhstan and Mongolia *over* the Himalayas to southern
India for the winter. George Lowe (1924–2013), the New Zealand-
born climber who supported Edmund Hillary and Tenzing
Norgay's ascent of Everest in 1953, said he had seen the geese flying over the top of Mount Everest – the peak is approximately
8000 metres – and, using GPS trackers, a team has recorded one
bird flying at 6000 metres.

---

* An imaginary line on the earth's surface cutting all meridians at the same angle, used
as the standard method of plotting a ship's course on a chart.

Bird brains

Most bird brains are tiny compared with primates', so how can they possibly carry out these computations? Not only are these brains tiny, they also lack a neocortex – the layered folded structure that covers the mammalian brain, in which are located the key brain regions for numerical processing in humans, monkeys and cats. It has been suggested that the pallium in the bird brain is functionally equivalent to the mammalian neocortex. Nevertheless, the neurons in their brains are more tightly packed than in mammalian brains. We now know that cockatoos (10 grams, 2 billion neurons) and corvids (1.5 billion neurons) have more neurons than many mammals and even some primates. Both the sulphur-crested cockatoo and the bushbaby, a primate, have brains that weigh about 10 grams, but the cockatoo has twice as many neurons as the bushbaby. The macaw's brain is no larger than a walnut, but it has more neurons in its pallium than the macaque, which has a brain the size of a lemon.[29]

Even the little budgie, with 150 million neurons, has more brain cells than mice, rats or even marmosets. What's more, given that the neurons in the brains of birds are more tightly packed than in mammals, this may make connections among them faster and easier.

Bird navigation, as I said, needs a map. Maybe the idea of a map in the brain seems ridiculous. What would it look like? Obviously not like a paper map folded and unfolded in the bird's head. Could it actually be made up of numbers? Think about Google Maps for a moment and how they are stored in a computer server, and then think about asking Google Maps for directions from your home to the local supermarket, or to Paris; it can provide both the location and the route, as a picture and as a sequence of commands. These maps are stored digitally: i.e. as a set of numbers, ultimately 0s and 1s. How is the route worked out? By calculations on the stored numbers.

I don't know how many memory elements Google needs to store its maps, but the humble homing pigeon has 690 million neurons, of which 437 million neurons are in the pallium. Given that each pallial neuron may be connected to hundreds, even thousands of other neurons, the storage capacity of the tiny pigeon brain is likely to exceed that of the storage of Google Maps needed for the homing routes. So the idea of a map in the brain of birds is not so ridiculous. If it is anything like Google Maps, then it really is carrying out arithmetical calculations on numbers.

Number neurons in the bird brain

The brain region involved in carrying out a Koehler task has been identified using single-cell recordings in the crow's pallium. This was a study by Andreas Nieder and Helen Ditz (see Figure 2 on p. 183). What they found was very similar to what Nieder had observed in monkeys, namely that there were neurons 'tuned' to specific numerosities – that is, one cell would respond after a short delay, most strongly to one object, another to two objects, another to three, another to four and another to five objects.

Ditz and Nieder conclude this study arguing for a deep evolutionary basis to our numerical abilities:

This finding could indicate that canonical endbrain [bird pallium and mammalian neocortex] microcircuits evolved in a common ancestor of mammals and birds. Perhaps this result might also provide a physiological explanation for the evolution of neuronal computations that give rise to numerical competence in both vertebrate groups. More comparative approaches in neuroscience will therefore be indispensable for deciphering these evolutionary stable neuronal mechanisms.[11]

Like humans and other mammals, birds have a hippocampus, and it is this brain structure that encodes space with specialized cells for the animal's location. This was originally discovered and elaborated by John O'Keefe at University College London and his colleagues.[30] O'Keefe won the Nobel Prize for this discovery along with May-Britt and Edvard Moser, who discovered – in the medial entorhinal cortex, a region of the brain adjacent to the hippocampus – 'grid cells' that provide the brain with a metrical internal coordinate system essential for navigation.

Evolution

The last common ancestor of mammals and birds existed about 300 million years ago. That's a long time even in evolutionary terms and much would have happened in that period. It is probably too early to tell whether the accumulator mechanism was present in the common ancestor, though we know that various timing genes and timing mechanisms have been conserved from even more distant common ancestors, so it would not be surprising if genes for building a counting mechanism have also been conserved.

In this chapter we have seen that birds with smaller brains

than mammals are, if anything, even better at counting and calculating. Some individuals are exceptional, such as Alex the
parrot, and some species, crows for example, are at least as good
as monkeys on the same tasks. Birds travel long distances to forage and to breed, and this requires sophisticated computation of
distance, bearing and time, in the same way that human navigators calculate their routes. So how do birds with their small brains
manage to achieve this level of mathematical excellence? It turns
out that these tiny brains pack the neurons together very tightly
so that many birds have more neurons than many mammals and
more even than some primates, and some of those neurons are
tuned to particular numerosities, just as in monkey brains.

CHAPTER 7

CAN AMPHIBIANS AND REPTILES COUNT?

Modern amphibians – mostly frogs, toads, newts, salamanders and caecilians (limbless snake-like creatures) – are a very ancient line. They were the first descendants of fish to inhabit the land about 350 million years ago in the Devonian period. Amphibians live in both water and land and typically lay their eggs in water. Like other ectotherms (cold-blooded creatures), amphibians have small brains (less than 0.1 grams), and a lower brain/body weight ratio[1]. Still, frogs have some 15 million brain cells, easily enough to implement a simple accumulator system.[2]

There are today more than 10,000 species of retile including lizards, snakes, turtles and crocodilians. Ancient reptiles are the ancestors of dinosaurs, not to mention birds, and ultimately, of us.

Despite the range and variety of amphibians and reptilians, there are fewer studies of their numerical abilities than some of the other animal groups I have described. Nevertheless, I'll show that counting in the wild is important and adaptive for both amphibian and reptilian species for foraging, and is absolutely critical for mating in some species.

Doing what comes naturally in the lab

The first lab study to investigate the numerical abilities of amphibians, in this case red-backed salamanders (*Plethodon cinereus*), was carried out by Claudia Uller and colleagues then at the University of Louisiana.[3] These creatures in the wild employ an 'optimal foraging strategy'. They like fruit flies (*Drosophila*), and when there are few flies they'll forage any size of fly, but when the flies are abundant they go for the larger flies.

Uller's pioneering study was simple and elegant, and followed a paradigm employed with monkeys that I described in Chapter 4. Monkeys spontaneously 'go for more' food when offered a choice.[4] Will these reptiles do the same?

Uller and her team offered the subject salamanders a set of choices: 1 vs 2, 2 vs 3, 3 vs 4 or 4 vs 6 fruit flies in separate experiments. The experimental set-up showed the live flies in glass tubes, so the subject salamander could see but couldn't eat the flies. The choice was evidenced when the salamander touched one of the tubes. It turned out that salamanders chose 2 over 1, 3 over 2, but neither of the other two conditions. There was, as Uller and colleagues concede, no way of determining whether the salamanders were responding to the total amount of 'fruit-fly stuff' or of fruit-fly movement, and salamanders are very sensitive to movement of small things in the environment – they could be edible.

Uller subsequently studied the salamander's ability to make large number discriminations (8 vs 12 or 8 vs 16). This time the stimuli were live crickets. Salamanders reliably chose the larger of 8 vs 16, that is, when the ratio between the sets was 1:2. They failed, however, at 8 vs 12, that is, a ratio of 2:3. Again, there was no control for movement or cricket-stuff. However, some salamanders spontaneously use the amount of fly stuff rather than numerosity.[5] It may be easier, in some circumstances, to go for more area or more volume than for more objects.

Calls of the wild

Numerosity is the critical information in the mating games of frogs. Female frogs use the vocalizations of males' 'advertisement calls', as a way of choosing a mate. In a playback experiment, Georg Klump and Carl Gerhardt, then at the University of Missouri, tested the kinds of calls the gravid female tree frog (*Hyla versicolor*) found particularly attractive.[6] The males of this species make calls with a particular pulse rate, and it is this rate that enables the female to ignore 'heterospecific' males and just focus on males of the same species. Of the compatible males, she prefers males with longer calls because longer calls correlate with the male's energetic investment in courtship and probably indicates his physical condition and fitness. One aspect of the call, perceived loudness, which may also indicate physical condition, appears to be less important; of course, this subjective loudness varies with distance, so is not a reliable indicator, certainly not as reliable as call duration. Pulse rate (number/duration) is here used by female to recognize males of the same species, but the number of pulses, distinct sounds in a call, could also signal a fit male. This of course requires counting, and there is a very beautiful example of just this.

Túngara frogs (*Physalaemus pustulosus*) are tiny – 1 to 1¼ inches long – and live in swamps in Central and South America. The frog breeds at night and feeds on small insects during the day. What is extraordinary about them, and relevant to our story, is that mating success depends on the numerical ability of both males and females.

During the breeding season, males display and females choose. Seem familiar? The females are seeking the fittest males. They often cannot see them but they can hear them making their 'advertisement calls', and neighbouring males compete to make the most attractive calls.

Gary Rose of the University of Utah, and many other scientists, have been studying these calls in this and several anuran (frog and toad) species for many years.[7] The male túngara call starts with a long downward sweep in sound frequency produced by an inflated vocal sac that it is almost as big as the frog himself. This has been called a 'whine' and has been described in the *New York Times* as sounding remarkably like a phaser weapon in *Star Trek*.[8] To compete with neighbors' calls, males add 'chucks' (brief, harmonically rich notes) at narrowly specified intervals to the end of their sweeps. Michael Ryan of the University of Texas and his collaborators have shown that females prefer males that produce complex calls with sweeps and chucks, but there is a reason why males don't always produce these calls. Frog-eating bats also prefer frogs producing complex calls, so attracting females comes with a degree of risk. It seems likely that males that can produce complex calls are larger and fitter than those that can't, making them attractive to both females and predators.

Still, the male has to compete with other males, and it does this by adding an extra chuck to exceed those of its neighbours.

That is, if it hears four chucks it will produce 4 + 1 chucks. The number of chucks per phrase is a good indicator of respiratory fitness, better than perceived loudness, for example, since this will vary with distance. It's actually a digital rather than an analogue signal and therefore retains more information over distance. There is an upper limit of the chucks that a male can make which is determined by its lung and vocal sac capacity. So the most attractive males can make about seven. This is energetically costly, especially if repeated, and risky. So both males and females count chucks.

It turns out that females prefer males that can produce more chucks, and this means that males must be able to count the chucks of their competitor, and add one or more to their own call, and that females must also be able to count in order to select the fittest male. If a female finds a male's call attractive, she will use the call, as well as ripples in the water caused by its production, to locate her new mate.

The túngara isn't the only numerical frog. Call matching is actually quite common among competing male frogs, and can be tested by playback experiments. In a study of the Australian quacking frog (*Crinia georgiana*), calls were broadcast from either of two loudspeakers and responses of a male were recorded. The test male accurately matched the number of notes in the playbacks. He really was matching numerosity rather than, say, total energy, since matching numerosity continued when the energy was varied.[9] It is not clear why these males should match calls from other frogs of the same species, 'conspecifics'. One hypothesis for call matching is that males are attempting to produce calls that are at least as attractive to females as those of rivals, without wasting energy.

Rose has discovered not only that numerosity is important for mate choice, but also how the frog's brain does it. There are neurons in the frog's auditory midbrain that select on the basis of the intervals between sound pulses – the rate of amplitude modulation. The pulse rate – that is, the interval between each sound pulse – distinguishes advertisement calls from other vocalizations. If the timing between chucks is off by just a fraction of a second, these neurons don't fire and the counting process does not start. Neurons in the auditory midbrain called the 'inferior colliculus' selectively respond only if a threshold number of sound pulses have occurred with the 'correct' timing; these 'interval-counting' neurons then count the pulses. It is not yet known whether the other species of frog that also seem to count sounds have the same type of mechanism in their inferior colliculus. Gary Rose has written to me that 'It will be interesting to know whether counting neurons exist that are able to count sound pulses that occur with really long intervals (e.g. 200 msecs between successive pulses), as in cases such as the Australian 'quacking' frog, and the Central American yellow treefrog' (personal communication).

Sensitivity to amplitude modulation (rate and number of sound pulses) in the auditory system is widespread not just in anurans but in vertebrates more generally, from fish to humans, using similar neural structures such as the inferior colliculus. Speaking just for myself, I find counting the number of sounds in rapid succession almost impossible. My human inferior colliculus is indeed inferior when it comes to counting. It may be that the human system, unlike the anuran system, needs to be trained, and I suspect that musicians can do this much better than me.

Frogs and toads begin their independent life as tadpoles. An individual tadpole will reduce the risk of being eaten by a

predator if it joins a group of other tadpoles, and will do so especially in the presence of a predator. How readily tadpoles will join
a group of conspecifics depends also on how sociable they are,
and some are more sociable than others. Here's a lovely study
by Giorgio Vallortigara with colleagues from the University of
Pavia, Italy, Alessandro Balestrieri, Andrea Gazzola, and Daniele
Pellitteri-Rosa, that takes advantage of this naturally occurring
behaviour.[10]

They tested two species of tadpole, green toad (*Bufotes balearicus*), which is normally very sociable and forms large stable social
groups, and the edible frog (*Pelophylax esculentus*), which is less
sociable, and typically forms only temporary groups.

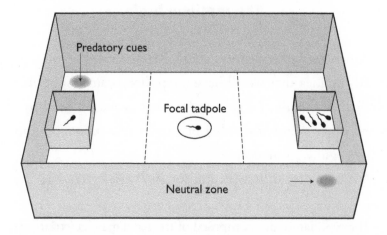

Figure 1. Testing the numerical ability of tadpoles. They join groups of
conspecifics especially when threatened. To test this, olfactory cues
from a predator are presented in one of the conditions. The numerical
question is whether the focal (subject) tadpole differentiate the groups
on the basis of numerosity. The sociable species, the green toad (*Bufotes
balearicus*), will select the larger group when the contrast is 1 vs 4 other
tadpoles. The less sociable species, the edible frog (*Pelophylax esculentus*), will choose the smaller group when not threatened.[10]

These tadpoles do not seem to have very good numerical abilities. This is different from fish we have tested: new-born fish are as good as adults in making numerical discriminations (see Chapter 8). We don't know whether these tadpoles will grow up to be better counters.

However, Vallortigara and his colleagues did test the adults of another species of frog, (*Bombina orientalis*).[11] This was also a study of spontaneously going for more food, but with proper controls for total mass of food (delicious larvae), surface area, volume and movement. These frogs choose the larger numerosity when the ratio is sufficiently large (1 vs 2, 2 vs 3, but not 3 vs 4, and 3 vs 6, 4 vs 8, but not 4 vs 6).

The reptilian brain

The American neuroscientist Paul D. MacLean (1913–2007) popularized the idea that we all have a reptilian brain responsible for primitive instinctive behaviour, but that evolution has provided us and other mammals with a neocortex that uniquely supports higher cognitive functions.[12]

According to Andrew E. Budson in *Psychology Today*:

The reptilian brain, composed of the basal ganglia (striatum) and brainstem, is involved with primitive drives related to thirst, hunger, sexuality, and territoriality, as well as habits and procedural memory (like putting your keys in the same place every day without thinking about it or riding a bike) . . . We all can make a choice, a choice as to whether we are going to give in to the primitive urges and desires of our reptilian brain or, instead, use our neocortex to control them.

In humans and mammals, it is this more modern, less 'primitive' neocortex that does the counting and computation. Since reptiles do not possess a neocortex that provides an implementation for higher cognitive functions, the question arises whether this primitive neural structure is capable of dealing with something as abstract as numbers. Here's a clue. The bird brain, which has evolved from the common ancestor of the modern reptiles 150–200 million years ago, also lacks a neocortex, and as we saw in Chapter 6, birds can be very good at number tasks.

Nevertheless, reptiles with their 'primitive' brains can manage quite complex behaviours. They can navigate mazes as well as birds or mammals. They can be social. Crocodiles incubate their eggs and guard the nest. Like some birds, some reptiles form pair bonds. Turtles demonstrate 'natal philopatry', navigating to and from a precise home beach often over thousands of kilometres, and remembering how to do this for the twenty years from when they leave to when they next return.

My friends and colleagues at the University of Padua, Maria Elena Miletto Petrazzini, Christian Agrillo and Angelo Bisazza, along with their colleagues at the University of Ferrara, tested the ruin lizard (*Podarcis sicula*) using a quantity control. The lizard could choose between two delicious larvae of different sizes, with ratios between the larvae masses of 0.25, 0.50, 0.67 or 0.75, and between two sets of larvae of the same size, with the same ratios, 1 vs 4, 2 vs 4, 2 vs 3 or 3 vs 4. The lizards spontaneously chose the larger mass, whatever the ratio, but failed to choose the larger numerosity, whatever the ratio.[5]

One theory, you may recall, is that animals only use numerosity information when making a choice as a 'last resort',[13] and bear in mind also that Koehler himself believed that animals could learn to use numerosity but didn't use it spontaneously. Maria Elena thought it might be worth seeing if these recalcitrant lizards could *learn* to use numerosity (see Figure 2).

In this study, the lizards were rewarded for selecting the larger disc (the area condition) or for the greater number of discs (the numerosity condition). In both conditions the ratios between sides were controlled in the same way – 0.25, 0.5, 0.67, 0.75

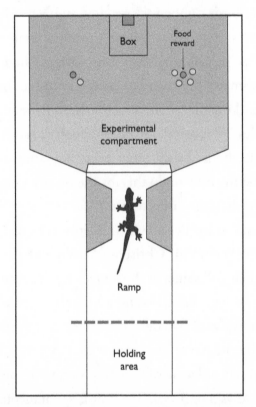

Figure 2. An experiment with ruin lizards. The lizard chooses between two sets of plastic discs. If it chooses the side with more discs, it gets a delicious larva as a reward. In a second condition, the lizard chooses between a larger and smaller disc and is rewarded if it picks the larger. The ratios between the sides in both the numerosity and area conditions are the same.[14]

ratios of areas, 1 vs 4, 2 vs 4, 2 vs 3 or 3 vs 4 in numerosities. So going for more should be equally difficult in both conditions. In this training set-up, it turned out that the lizards were actually more accurate in the numerosity condition than the area condition. For example, the lizards were better able to distinguish between 3 vs 4 discs than they were between 2 discs that had relative areas of 0.75 (different areas).

So, yes, they can learn to use numerosity, and are actually better at it than using area.[14] Incidentally, so are human infants. The question is: why don't they use this ability spontaneously? As I have mentioned several times, in real life numerosity and area typically co-vary, and sometimes it is easier to go for more area than for more objects, and sometimes not.

The numerical ability of tortoises too came under the careful scrutiny of Vallortigara and colleagues.[15] They used what should now be a familiar test of whether they would 'go for more' items of food, and whether their performance would reflect Weber's Law. The subjects in this study were Hermann's tortoises (*Testudo hermanni*). The question was whether they would go for more tomatoes (numerosity condition), and whether they would go for more tomato stuff when the size of the tomato slice was varied (area condition), in a two-choice task. The tortoises were confronted with the same ratios 0.25, 0.50, 0.67 and 0.75 in the sizes of the tomato slices and in the number of tomato pieces – 1 vs 4, 2 vs 4, 2 vs 3 and 3 vs 4.

The results were clear. The tortoises were pretty good in both conditions, and in both they were better for the 0.25 (1 vs 4 pieces) and 0.50 (2 vs 4). So, yes, they were able to count tomato slices, though it was also the case that since the slices were all the same size, the total area of tomato varied with numerosity. Again the issue arises as to whether the tortoises were really using numerosity or just the total amount of tomato stuff. Each tortoise was tested in both conditions. If they were using the amount of stuff to do the numerosity task, then it might be expected that performance by each subject on each ratio would be correlated. But that turned out not to be the case.

These are very simple computations that the tortoise has to learn in the lab, but far more complex computations are employed in the wild. Like migrating birds, one of the tortoise's cousins, the aquatic turtle, migrates and therefore needs to plot routes to and from its breeding and foraging areas.

Turtle navigation

Charles Darwin was impressed, and puzzled, by turtle navigation:

> Even if we grant to animals a sense of the points of the compass, of which there is no evidence, how can we account, for instance, for the turtles which formerly congregated in multitudes, only at one season of the year, on the shores of the Isle of Ascension, finding their way to that speck of land in the midst of the great Atlantic Ocean?[16]

We now have evidence that turtles do have 'a sense of the points of the compass'.

Loggerhead turtles (*Caretta caretta*) are large, with adults weighing 135 kilos (300 lb), with the largest specimens as much as 450 kilos (1000 lb), and they are very long-lived, indeed some of the longest-lived animals, reaching sexual maturity at 17–33 years, and could have a lifespan of 70 years. A hatchling will return to the natal beach when it is sexually mature – that is, after 17 or more years. This means, of course, that it has to remember this location and how to get back there after an extraordinarily long time. So it needs very good long-term memory as starters. How it remembers the location and the route back is beginning to be understood.

It is now known that the turtles use the earth's magnetic field, as do some long-distance birds, by detecting the intensity and the angle at which field lines intersect earth's surface. The angle is ninety degrees at the magnetic poles and zero degrees at the magnetic equator. This means that there are 'isolines' of angle and intensity – like isobars – across the globe that could provide the turtle with a unique magnetic signature of its natal beach. The juvenile would need to remember two key parameters – its precise intensity and the angle of the magnetic field – plus it would help to note how these change as the turtle swims away from it. Finding its way back home would be a case of reversing the change for these two parameters. There are two problems with this route-finding strategy. First, the earth's magnetic field changes in a systematic way, so using this strategy means that the turtle would end up in not quite the same place as it started out. However, this provided the investigators with a perfect opportunity to test the hypothesis of the magnetic map. If there is a systematic and predictable change in the turtle's return route, this would be evidence that indeed they are using the earth's magnetic field.

The exact location may have moved along the coast, or inland, or out to sea.[17] Using a nineteen-year history of returning loggerheads, this is exactly what two of the great experts on turtle navigation, Rogers Brothers and Kenneth Lohmann, from the

University of North Carolina, found.[18] The returning turtles ended up not exactly where they started but in a location predicted by the shift in the earth's magnetic field. It's a kind of numerical imprinting.

As with bird navigation, if we think of the parameters of the loggerhead turtle's geomagnetic map as numerical, like any digital map, then turtle navigation could involve memorizing numerical parameters of magnetic intensity and angle, along with a sample of the rate of change of these parameters in order to locate its current position and to find its way home. These memories would have to be very long-term indeed, lasting twenty or more years.

So to sum up what we know: reptiles in the lab but without training have been shown to be capable of spontaneous numerosity comparison, at least where food is concerned, but only up to four, and for larger numerosities when the Weber fraction (ratio) is 1:2, and it is reasonable to suppose that this ability is used in foraging in the wild. In the wild, frogs count croaks in mating games, and several species, including the male túngara, can count the number of notes in the calls of a neighbour; they progressively add notes to their own calls to 'one-up' their competitor, thereby making them more attractive to gravid females.

These capabilities can be readily explained with the simple accumulator system outlined in Chapter 1. As I have suggested, this system is small and simple, and actually doesn't need a neocortex to implement it. For this reason, the brain has the neuronal capacity to implement several accumulator systems.

In reptiles, food items identified by the selector increment the accumulator by a fixed amount, and the reference and working memory components enable comparison between two sets of food items. The selector in the frog would need to be set to increment by note or 'chuck', and one neuron, which also selects on the basis of rate (duration/number), accumulates the number of consecutive notes that occur with correct timing. Gary Rose, the frog expert, told me that 'we currently do not know how the counting information, which is sensory, is then translated into the motor action of matching the number of call notes ("chucks" in the case of *Physalaemus pustulosus*)'.

CHAPTER 8

CAN FISH COUNT?

We have seen that our near and distant mammal relatives have good or very good counting abilities. It is also important for the safety of newly hatched chicks and other birds to be able to count and to carry out calculations. Amphibians and reptiles, too, count when foraging or when choosing a mate. Apart from cetaceans with their enormous and complex brains, these are all land-dwelling creatures. Here we consider whether fish, whose environment is so different from our own, can count, and why they might need to.

Fish comprise more than half of all extant vertebrate species,[1] so it is perhaps surprising that so little attention has been paid to their numerical abilities until very recently.

In 2017, out of forty-eight studies only twenty-three controlled for other continuous non-numerical cues, such as total area, and some fish species were more popular than others: eight were on guppies (*Poecilia reticulata*), but only two were on zebrafish (*Dario rerio*) (about which more later).[1]

Given the relatively smaller brains of fish, one might have expected that they have poorer cognitive abilities in general than 'higher' vertebrates – reptiles, birds and mammals. Actually, some have better memory abilities both in the wild and in the laboratory. Famously, many species, such as salmon, can remember the properties of the river in which they spawned for several years and successfully return for mating. The route through a maze can be remembered three months later.[2]

Some of the most complex fish behaviour is instinctive and relies very little on any kind of higher cognitive ability. I had the privilege of being taught as a student by Niko Tinbergen (1907–1988), who went on to win the Nobel Prize in 1973. I wasn't really interested in the behaviour of three-spined sticklebacks (or herring gulls, his other special subject), but I find that I can still remember the lectures very clearly, which is more than I can say for the other lectures I attended. So he must have been a great teacher as well as a great scientist. All the other lecturers wore a gown and a tie. Those were the days! But not Tinbergen. I still remember the key theoretical terms: *sign stimulus, innate releasing mechanism,* and the curious and complicated *fixed action pattern* of the three-spined stickleback (*Gasterosteus aculeatus*). Clearly, a decent-sized brain is needed for this. Here's how Tinbergen himself described it:

> In nature sticklebacks mate in early spring in shallow fresh waters. The mating cycle follows an unvarying ritual, which can be seen equally well in the natural habitat or in our tanks. First each male leaves the school of fish and stakes out a territory for itself, from which it will drive any intruder, male or female. Then it builds a nest. It digs a shallow pit in the sand bottom, carrying the sand away mouthful by mouthful. When this depression is about two inches square, it piles in a heap of weeds, preferably

thread algae, coats the material with a sticky substance from its kidneys and shapes the weedy mass into a mound with its snout. It then bores a tunnel in the mound by wriggling through it. The tunnel, slightly shorter than an adult fish, is the nest.

Having finished the nest, the male suddenly changes color. Its normally inconspicuous gray coloring had already begun to show a faint pink blush on the chin and a greenish gloss on the back and in the eyes. Now the pink becomes a bright red and the back turns a bluish white.

In this colorful, conspicuous dress the male at once begins to court females. They, in the meantime, have also become ready to mate: their bodies have grown shiny and bulky with 50 to 100 large eggs. Whenever a female enters the male's territory, he swims toward her in a series of zigzags – first a sideways turn away from her, then a quick movement toward her. After each advance the male stops for an instant and then performs another zigzag. This dance continues until the female takes notice and swims toward the male in a curious head-up posture. He then turns and swims rapidly toward the nest, and she follows. At the nest the male makes a series of rapid thrusts with his snout into the entrance. He turns on his side as he does so and raises his dorsal spines toward his mate. Thereupon, with a few strong tail beats, she enters the nest and rests there, her head sticking out from one end and her tail from the other. The male now prods her tail base with rhythmic thrusts, and this causes her to lay her eggs. The whole courtship and egg-laying ritual takes only about one minute. As soon as she has laid her eggs, the female slips out of the nest. The male then glides in quickly to fertilize the clutch. After that he chases the female away and goes looking for another partner.[3]

This sequence is a *fixed action pattern*, and the *sign stimulus* is the gravid female. For her, the sign stimuli are the zigzag dance and the male's red breast, which activate the *innate releasing mechanism* and the sequence of actions begins and typically goes all the way through to completion. But the idea that fish, with their small brains (10 million neurons), could deal with concepts as abstract as number seemed outlandish.

Thus, Angelo Bisazza at Padua University in Italy, one of the pioneers of fish research, confessed initially he was reluctant to test their numerical abilities. He wrote:

> In the late eighties I reasoned with one of my co-authors, Guglielmo Marin, about the possibility that fish could possess counting abilities. At that time I was a behavioural ecologist and I was familiar with the fact that in many species (e.g. guppies and peacocks) females choose males based on the number of colour spots of the prospective mates.
>
> We devised some possible ways to test these hypotheses. However, at that time, fish were believed to have very primitive cognitive abilities . . . Studying mathematical abilities of fish seemed a bit bizarre at the time and we were not brave enough to pursue such a high-risk project.

A brilliant student, Christian Agrillo, applied to work with Angelo, hoping to be able to study monkeys, but instead found himself working on fish:

> In 2003 there was no study on the numerical abilities in fish, probably nobody was so crazy to waste time on it. Angelo and I were crazy enough to make this bet, to spend the whole PhD period to establish whether fish can count. That's the way everything started ... the world really needs two high-sensation seekers like us to see whether numbers can exist also in the underwater world!
>
> Luckily enough, everything worked quite well. When we published the first extensive paper about it in 2008,[4] most of the important media covered the news (BBC, CNN, *National Geographic*, RAI).

In addition to this study, two things have prompted the recent work on the numerical abilities of fish: first, reflections on the shoaling behaviour of small fish; and secondly, an extraordinary event that took place over 300 million years ago.

Shoaling

It's been known for many decades that joining a group, a shoal or school can be beneficial. It will be easier to find a mate in a group. Species that feed on large particles improve their chances of finding them with many eyes looking; and being in a large group reduces the risk of an individual being eaten by a predator. The larger the shoal, the better it will be for reproduction, feeding and safety. Therefore, it would be advantageous for fish to be able to choose the larger shoal.

One of the earliest demonstrations of the possible use of numerical information in shoal choice was carried out with minnows (*Pimephales promelas*), sometimes in the presence of a predator, the largemouth bass (*Micropterus salmoides*). Each minnow was offered a choice of two shoals on opposite sides of the tank. The shoals ranged from one to twenty-eight minnows, and the test minnows chose the larger shoal right across the number range, with or without the presence of the predator. This suggests that joining the larger shoal is instinctive, but depends on the ability to assess the numerosities of the two shoals.[5] Now, although numerosities were manipulated by the experimenters, it wasn't clear that the test fish were responding to the numbers or to shoal density, since the different-sized shoals occupied the same tank volume.

Our friend the three-spined stickleback, when not in the mating season, also shoals. Like many other shoaling fish, it does respond to the density of the shoal when making a choice, since numerosity and density typically go together in real life. With the same numerosity of the two shoals, the stickleback will prefer the denser, but with the same density, it will prefer the more numerous.[6] The standard set-up for recent studies of spontaneous – no training – numerical discrimination is shown in Figure 1.

Studies of many species of shoaling fish have shown that fish choose the bigger shoal, and it's easy to vary experimentally the numbers on each side, so that one can calibrate the ability to estimate or compare numerosities.

Here's one study that I did with my friends at Padua University using the apparatus shown in Figure 1 (overleaf). The subjects in our experiment were guppies (*Poecilia reticula*). We had a particular objective in this study to see if these small fish had two numerosity recognition systems, as it was claimed for other vertebrates including

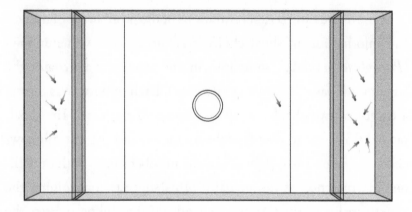

Figure 1. Three-tank method. The test fish is in the central transparent tank and the shoals are in the transparent side tanks. Which side will the fish go to? No training is needed: the fish will spontaneously choose the side with more fish.[7]

us humans: one small numerosity system and one for larger numerosities. As I noted in other chapters, the small numerosity system, sometimes called the 'subitizing' system for numerosities ≤4 (see Chapter 2), has two interesting characteristics. First, it is virtually error-free, and in humans, very fast. Second, when comparing two numerosities ≤4 there is no *ratio effect* (see Chapter 2). That is, it is just as easy to select the larger set with four objects compared with three objects, as compared with one object. For numerosities greater than four, the ratio effect (Weber's Law, see Chapter 1) kicks in, so that comparing nine with five objects is more accurate and faster than comparing nine with eight objects.

We first tested this hypothesis with Italian students. We didn't, of course, immerse them in tanks of water, nor ask them to compare shoals of fish, but to select the larger of two arrays of dots presented successively. We measured their accuracy and the speed of these judgements.[7]

We found what many other studies have reported, namely that for small numerosities there was no effect of the ratio between the two arrays on either accuracy or speed, while for larger numerosities there was a ratio effect for both of these measures. In humans, the brain processes large and small numerosities differently.[8]

Are these two systems present in the guppy brain?

It turns out that they are. What is more, the two systems are present at birth. We tested 100 one-day-old fish and 140 'experienced subjects'. Here are the numerosities and the ratios we used:

Ratios	0.25	0.33	0.50	0.67	0.75
Small numerosities	1 vs 4	1 vs 3	1 vs 2	2 vs 3	3 vs 4
Large numerosities	4 vs 16	4 vs 12	4 vs 8	4 vs 6	6 vs 8

The day-old fish performed identically to the adults. This suggests that the two systems are wired in and start functioning without the benefit of experience.[7]

A variant invented by my friends at Padua University allows the test fish to see only one fish at a time (see Figure 2, overleaf). The Padua team used mosquitofish (*Gambusia holbrooki*), a small freshwater shoaling fish.[9] The baffles in the tank meant that the test fish could swim about freely, but could see only one fish at a time. These little fish were able to select the larger shoal when the contrast was in the small number range (3 vs 2) and also in the large number range (4 vs 8). This means the test fish had to sum the number of fish on each side of the tank, remember the sums, and carry out a numerical comparison between the two shoals in order to choose the larger shoal.

Can fish choose the more numerous set if the objects aren't other fish but just random arrays of dots, like monkeys, for example, are able to do (see Chapter 4)? Several studies from Padua

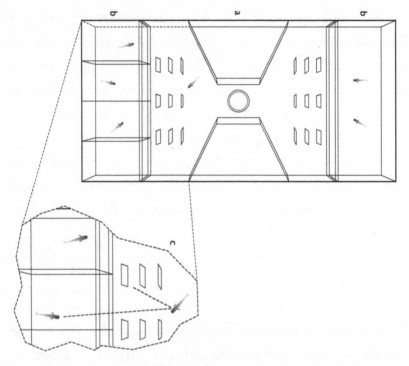

Figure 2. One at a time. The test fish can swim about freely in its own tank but can only see one fish at a time in each shoal, and only one shoal at a time, but can choose the larger shoal in contrast between three and two and between eight and four.[9]

show this. Here's one example that I was involved with, and which I'll say more about later because it tells us something interesting about two (fish) heads being better than one. We offered guppies two arrays of dots and rewarded them with food if they chose the array with more dots in the apparatus shown in Figure 3. The dot arrays were controlled for non-numerical cues such as surface area, density and the overall space occupied by the dots. Fish were initially trained on easy contrast with a 1:2 ratio (5 vs 10 or 6 vs 12 dots), and then tested on the more difficult 2:3 ratio (8 vs 12 dots) and 3:4 ratio (9 vs 12). Half the fish were trained to choose the larger array and half the smaller array.

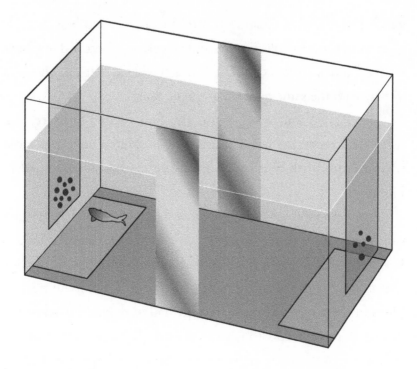

Figure 3. The apparatus used to train guppies to choose the more numerous (or less numerous) array of dots. A food reward would be given near the array with more dots (or fewer, depending on the experimental condition).[10]

We found that the fish were able to manage the 2:3 ratio, showing that they were able to represent these numerosities and learn to choose the larger, or smaller, of the numerosities. These were fish working alone. In pairs, they did better. See below.

Exactly three, exactly four, exactly more

Experiments with dot arrays also enable us to see if fish can represent particular numerosities, rather than just discriminate on the basis of relative numerosities. The basic idea is called 'match-to-sample', which I described in Chapter 1, with the original and best example given with Otto Koehler's crows and ravens, as I described in Chapter 6 on birds. In this case, the fish is rewarded if it can choose a display with the same numerosity as the sample.

The Padua team were again the first to demonstrate that fish could represent exact numerosities. Here's an experiment with mosquitofish that shows this. The method is presented in Figure 4.

In a comparable study, the same team showed that guppies can represent fourness.[12] In fact, they used a very similar task to Koehler's. If you recall, the bird had to remove the lid of the box that displayed the right number of blobs on it – the number of blobs in the sample. Here, the guppy had to move a lid below the display with four dots to get the reward. The alternatives were 4 vs 1, 4 vs 2, 4 vs 8, 4 vs 10, so just choosing the larger (or smaller) numerosity wouldn't work. The fish had to be able to represent the exact number, four.

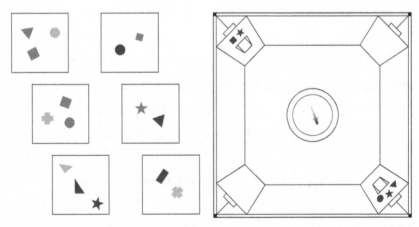

Figure 4. Sample dot arrays are given on the left, and the tank on the right. If the fish chooses the door with trained numerosity above it, then it can swim out and join its friends.[11]

Good fish, bad fish and the origins of dyscalculia

In our study of pairs of guppies (dyads) vs singletons, we noticed that some fish when tested alone seemed to be better at doing the number tasks than others.[10] In previous studies, the variation in performance was treated simply as statistical noise. What was regarded as important was the 'central tendency' of the whole group. Did the group on average show the desired effect? Of course, there could be all sorts of reasons for variation in fish performance, just as there might be for us. The normal statistical tests that are applied not just in fish studies, but more generally, can be used to see whether the test fish *on average* choose the correct (the rewarded) numerosity more often than would be expected by chance. In the case of our fish, yes, they did. But what about the fish that didn't choose the correct numerosity? Were they just having a bad day; were they just not interested in the task; or were they maybe just tired? If we tested them again, would they still fail to choose the correct numerosity?

Since we were interested in comparing the performance of dyads and singletons, we had recorded which fish were good and which were not so good, so that when we put them into dyads, we could see whether they improved. The reason we thought that this might happen comes from an earlier study of a similar numerosity task we had carried out with human adults.[13] The two participants would see two sets of dots briefly and had to decide first individually which set was more numerous. If they disagreed, then they would discuss it and come to a joint decision. In that study, we found that humans were better when they discussed their differences than if we took the average of their choices made alone. We had a fancy explanation for it that we called 'weighted confidence sharing': that is, the judgement made by the pair would reflect the confidence of the two participants. Thus, for a given trial, where the two participants disagreed and they discussed what should be their joint decision, they would typically go for the judgement of the more confident. This, we thought, was because confidence itself would reflect how accurately the respective brains perceived the stimulus. Of course, there are a lot of assumptions built into this line of reasoning, but it turned out to be justified by the evidence.

I will discuss in the next section whether dyads were better than singletons, and why. For this section, the important point is that when we re-tested the bad singletons when they were members of the dyad, most of them were still bad. So there really did seem to be individual differences in the numerical abilities of fish, and this was perhaps the first time individual differences have been highlighted in animal studies. Why should some individuals be better than others?

We don't have an answer at the moment, but, as with us humans, one source of difference could be genetic. Currently, we are testing another species of small shoaling fish called the zebrafish on account of its stripes. We chose this species because its genome has been sequenced and we know how to manipulate the zebrafish genome to create families with specific genetic properties. This means we can test candidate genes we find in humans. That is, if we find that a variant of gene X is a bit more frequent in humans with poor numeracy, is this difference really the cause? So we create a family of zebrafish with this variant and see whether they are also poorer on a numerical task.

We know that about 4–5 per cent of people have serious

difficulties with numbers and arithmetic, which they cannot learn in the usual way. This is usually called 'dyscalculia', and like the much better-known condition dyslexia, it is present from birth (or the earliest age at which it can be tested) and persists into adulthood and old age. The root cause, what I have called the 'core deficit', is in representing the numerosity of sets of objects, and sets and operations on sets are the basis of understanding number and arithmetic. That is, people with dyscalculia are bad at the kinds of tasks we give to fish, comparing and identifying numerosities. Evidence from twins suggests that there is a genetic component to the core deficit in many cases, but we have yet to identify the gene or genes involved. I have reviewed the evidence about the causes of dyscalculia in a recent book.[14]

This is where the zebrafish come in. We can't create human families with a particular gene variant, though we might, with luck, find them, but we can, as I said, create zebrafish families with the gene variant.

Meritocratic leadership

Small fish are prey to bigger fish – minnows for the largemouth bass, for example – so they have evolved to join shoals, the larger the better. This means they need to be able to estimate shoal sizes and choose one on the basis of its numerosity. But it's not just a matter of joining. Since shoal members are swimming around all the time, the shoal must move around as a group, so that individuals are not isolated and can be picked off by a predator. That is, they must all decide to move in the same direction and at the same speed. How do they make this decision?

The most popular model of group advantage in decision-making in humans and in animals has been called the 'many wrongs' (MW) hypothesis. According to this, each individual makes an estimate that is an approximation to the correct one but with some error; then, if these errors are randomly distributed around the true mean, they will cancel each other and the whole crowd will be more accurate than most, if not all, its single members.[10] This hypothesis has been advanced as an account of navigation by flocks of birds.

In the minimal social group, a dyad, MW predicts that the group accuracy will be the average of its members. Is this true or is there another model that explains group advantage better? We have called a second model 'meritocratic leadership' (ML) and argue that it applies 'if some members are more accurate than others to accomplish the task. In this scenario the group would enjoy an advantage provided that collective decisions are guided by its best performing members.'[10]

This kind of mechanism is thought to be at the basis of collective decision-making in honeybees, where one or a few informed individuals can determine the decision of the whole group (see Chapter 9). We therefore had two very clear predictions to test which model was correct. If MW was correct, then the accuracy of the dyad should be the average (the mean) of the two fish acting alone; if ML was correct, then the dyad should be as accurate as the better of the two fish.

The next study was the result of two slices of scientific luck, or perhaps chutzpah on my part. First, there was my colleague Bahador Bahrami, with whom I had worked previously. He had just published in *Science* a brilliant study titled 'Optimally interacting minds' in which two human individuals, discussing a perceptual judgement where they had previously disagreed, were more accurate than each making a judgement alone, as I mentioned earlier. So I asked Bahador my standard question: have you tried this with numbers? If we found the same kind of result then the minds would interact optimally using weighted confidence sharing in a cognitive judgement as well as the perceptual one. It turned out that we found very similar results when our subjects had to choose the more numerous of two dot arrays.[13]

When I visited Padua later that year, I had a chat with Angelo Bisazza and Christian Agrillo. Now I asked them my new standard question: have you tried this with fish? We discussed it for a bit, and decided that here was a way of testing whether two fish heads, like two human heads, are better than one in making a decision, and the theoretically important question of whether, if there was a collective advantage in fish, this was because errors of each were averaged (MW) or due to the influence of the more able fish (ML).

In our first experiment, we used a set-up like the one shown in Figure 1 (p. 220), and tested how the singletons and the dyads performed on shoal six vs shoal four, which is close to the limit of guppy numerical ability. The dyads were significantly better than the singletons. So there is a collective advantage in choosing the larger shoal. We also found that the dyads were significantly better than the average of the two fish tested separately. In fact, the performance of the dyad was the same as the performance of the better fish. This is evidence in favour of ML. We were using shoals of other guppies, and it may be that the collective advantage is specific to that situation, but may not reflect a general advantage in numerical tasks. To test this, we trained thirty more guppies to discriminate between two numerosities by rewarding the choice of the greater numerosity, as shown in Figure 3 on p. 223. Again we found, first, that indeed, dyads were better at choosing the greater numerosity; and second, that the dyads were able to choose correctly when the ratio was 3:4, which was beyond the average capability of the average single fish! And again, the performance of the dyad was determined by the capability of the better fish.

Thus dyad performance is determined by the better member taking a leadership role, and hence in favour of what we have called meritocratic leadership and against the average of errors (MW).

We knew that leadership can emerge spontaneously in the shoaling behaviour of fish from studies of shoal foraging behaviour. Recently, additional light has been thrown on this problem in a lab study of damselfish (*Dascyllus aruanus*) that live in lagoons in the Great Barrier Reef, Australia.[15] It turns out that these shoals have leaders. They are not the biggest or the most dominant, but rather the most active, since they are most likely to initiate a movement that others follow when the shoal is ready to move.

How good are fish?

A popular way of testing numerical abilities in other species is to identify how good they are at choosing the larger (or sometimes the smaller) numerosity, usually where there are two options. So how do fish compare? They are not as good as humans or apes (which are capable of up to 9 vs 10 discrimination), and monkeys can sometimes achieve 7 vs 8 (see Chapter 4). But they are at least as good as and sometimes better than other mammals (e.g. dogs, 6 vs 8; horses, 2 vs 3 [see Chapter 5]) and birds (pigeons, 6 vs 7; domestic chicks, 2 vs 3 [see Chapter 6]).

Laboratory studies based on the observation of spontaneous behaviour and on training procedures have shown that the accuracy of fish numerosity comparison tasks equals that of many birds and mammals. As we have seen, mosquitofish can discriminate up to a 0.67 ratio, but not a 0.75 ratio (e.g. 8 vs 12, but not 9 vs 12), while guppies can discriminate up to a 0.75 or even a 0.8 ratio in a training study. Three-spined sticklebacks can even discriminate 6 vs 7 (0.86). This compares very favourably with non-primate mammals, as we saw in Chapter 5.

Three hundred and fifty million years ago

Three hundred and fifty million years ago something very extraordinary, and somewhat mysterious, happened to the ancestor of teleost (ray-finned) fish, by far the largest group of vertebrates. Their whole genome became duplicated – that is, each gene had a duplicate, a redundant 'paralogue'. As my colleagues Christian Agrillo and Angelo Bisazza put it, 'Gene duplication is recognized as a major force in evolution because the duplicated copy, released from its original function, can be the origin of a new function. The duplication of the entire genome offered to teleost fish an enormous evolutionary and adaptive potential early in their evolutionary history.'[1] The redundant genes can provide the genetic raw material for evolutionary innovation because they are released from selective pressure, and could gain new functions. Another possibility is that by having two genes now doing the job normally done by one, together they could produce a double dose of the relevant protein.

The families of genes that have enriched functions as a result of duplication 'are related to ion channel and transporter activity. Ion transport needs to be tightly regulated in any cell. However, neurons are the cells that most strongly rely on a repertoire of diverse ion channels and transporters ... also studies of protein families in zebrafish have shown that genes involved in neuronal function have often retained both paralogues.'[16] That is, duplication could modify neuronal functioning. Another gene family, important in brain development in zebrafish, has also been enlarged and enriched by duplication.[17]

Agrillo and Bisazza sum up the potential cognitive advantages of duplication thus:

> Recent analysis found that, in modern fish, the retention rate of genes implicated in cognitive processes is much higher than the average retention rate of the rest of the genome, suggesting that 'cognition genes' have been the target of selection early in the history of fish and that new cognitive abilities may have played a role in the evolutionary success of this group.[1]

Fish brains

Is it possible to identify the brain mechanism that fish use to make numerical decisions? There is no point in putting a fish in a big functional magnetic imaging machine: the brains are far too small, below the resolution of the machine, to pick out particular tiny regions doing the numerical work.

One way pioneered by Scott Fraser and his team at the University of Southern California working with my colleagues Giorgio Vallortigara and Caroline Brennan is to see if the brains of zebrafish are changed by the fish seeing a change of numerosity.

In this study, the dots changed in individual size, position, surface area and density, while maintaining their numerosity. Then the fish would see a different sets of dots with a different numerosity but with the same overall surface, size and variations of position. There was also a control group, zebrafish faced where the numerosity of the set of dots did not change. So the experimental question was how did the brains of the zebrafish react when the numerosity changed? Did they even notice it?

It turned out that the brains were changed in the fish by a change of number. In particular the change occurred in a region of the pallium, the most evolutionarily advanced part of the brain and it has been claimed for bird brains to be functionally similar to mammalian neocortex, which is the structure that includes the hub for number processing in primates and other mammals.[18]

So, can fish count?

The great Cambridge ethologist, W. H. Thorpe (1902–1986), reviewed the ability to count in many bird and mammal species which he defined as the ability 'to abstract the concept of numerical identity from groups of up to seven objects of totally different and unfamiliar appearance'.[19] This was essentially the definition I adopted in Chapter 1. In 1962, the date of Thorpe's review, there was no research on the numerical ability of fish, and so he had no opportunity to test whether fish could meet his definition.

We now know that fish can certainly meet this definition. They can learn to match the numerosity of abstract shapes, as in the studies shown in Figures 2 (p. 222) and 3 (p. 223). And they spontaneously choose the larger shoal of fish when the shoals are swimming around freely and do not present a constant visual image. In fact, when the experiment is designed in such a way that the test fish can only see one fish at a time, as shown in Figure 2 (p. 222), it can count and sum the fish on each side of the tank and then choose the side with more fish with considerable accuracy.

We are also beginning to see the brain mechanism in the fish pallium that does the counting. Further research may be able to see the mechanism counting in real time.

Like us, not all fish are equally good at counting. This is not just the difference between species – in my experience, guppies seem to be better in general than zebrafish – but even within a species there are individual differences. To the extent that these differences are genetic, our current experiments may in due course provide a model for testing the genetic basis of the awful handicap of dyscalculia in humans.

This numerical ability is importantly adaptive because it enables fish in the wild to choose the safety of the largest available shoal. Here, individual differences play a role. The fish with the best numerical ability will lead other fish, and perhaps their leadership role ensures that the shoal stays together by heading in the same direction. It may turn out that the genes that make an individual fish good or bad at numbers are the same genes that predispose us in the same way.

CHAPTER 9

ARE BIGGER BRAINS REALLY BETTER?

In previous chapters I described the numerical abilities of vertebrates, all of which are rather like us in many ways. They have backbones and endoskeletons, and are typically roughly bilaterally symmetrical, even their brains. Although only mammals have a neocortex, fish, birds and reptiles have brain structures that seem to do similar jobs to our own. Of course, their brains are smaller and less complex (though whales' and dolphins' brains can be much bigger and in some ways more complex). It turns out that the most extraordinary counter is not the one with the biggest brain, but the species with one of the very smallest, the ant, in particular the *Cataglyphis fortis*, a species that lives in the searing hot desert of North Africa.

Invertebrates are an enormous group of animals – insects, spiders, squid and octopus, among many, many others. Typically, they have tiny brains, apart from cephalopods such as the octopus and cuttlefish. It turns out that these creatures can count, and indeed may be very good at counting.

Charles Darwin in *The Descent of Man* noted that tiny brains can do a lot:

> It is certain that there may be extraordinary activity with an extremely small absolute mass of nervous matter; thus the wonderfully diversified instincts, mental powers, and affections of ants are notorious, yet their cerebral ganglia are not so large as the quarter of a small pin's head. Under this point of view, the brain of an ant is one of the most marvellous atoms of matter in the world, perhaps more so than the brain of man. (1871 edition, p. 54)

The brains themselves, though tiny, may be more sophisticated than at first imagined, and not just a tiny and incomplete version of our own. A century ago, the great neuroanatomist Santiago Ramón y Cajal (1852–1934, Nobel Prize 1906) produced pioneering – and beautiful – studies of the nervous systems of insects. He likened their neuroanatomy to a 'fine pocket watch', as opposed to the 'rough grandfather clock' of the vertebrate brain.

Arthropods, animals without a backbone, with a segmented exoskeleton instead of an endoskeleton, diverged from our evolutionary line at the time of the Cambrian explosion 600 million years ago. Naturally they live a very different kind of life from us vertebrates, but they live in the same world and need, as we do, to be able to read the language of the universe if they are to survive and prosper. They have to navigate, forage efficiently, build or find habitations, and reproduce.

However, they have very different brains from ours – much smaller, of course, and differently constructed. As I have mentioned before, brains are metabolically expensive. Our brains take up about 2.5 per cent of our body weight but burn up more than 15 per cent of our basal metabolic energy. In insects they can reach over 8 per cent of body weight and even 15 per cent in some of the smallest ants,[1] so one might expect there to be some costs to this. William Eberhard and William Wcislo of the Smithsonian Tropical Research Institute in Panama, in a really fascinating review, conclude that while one might expect very small brains would lead to lifestyles that are 'behaviorally less demanding ... at least some small-bodied animals express the same kinds of behavior as their large-bodied relatives'.[2]

Nevertheless, in this chapter we will ask if these creatures with really tiny brains understand at least one aspect of the language of their universe. As I argued in Chapter 1, the main difference among species is not whether they can count, since the basic accumulator mechanism is very simple, but *what* they can count, and whether they can generalize from the count of one type of thing to another type of thing. And these abilities may depend on many and 'higher' cognitive abilities. But is this true of invertebrates? One other thing: for an activity to be classified as counting, the result of that activity has to be able to enter into computations.

Computers are not necessarily better simply because they are bigger. My first experience was of an IBM 360 mainframe computer. Although I was programming it, I wasn't allowed into the room it filled. That was strictly reserved for the 'operators'. It must have weighed at least 2000 kilos, and had a memory capacity of 64 *kilo*bytes. Compare this with the tiny Raspberry Pi 4 that was created to help kids learn coding. It has 4 *giga*bytes of memory, 60,000 times more, and weighs just 23 grams. And of course the Pi is many times faster.

Bees

Honeybees have been the most intensively studied from a numerical perspective. Now bees have really tiny brains, about one *millimeter* cubed, with fewer than 1 million neurons, which is, at least, four times as big as the ant brain. But are bigger brains better when it comes to counting?

You might expect that having a tiny brain means that honeybees will be very limited in what they can do. But this is not the case. The worker bee has to locate and build a hive with beautiful and symmetrical hexagonal honeycombs. She can collect pollen and nectar from those plants that produce them; she can sting and avoid being stung. Bees are also tidy creatures that clean the nest of debris, and of course are very sociable and community-minded.

When bees find food, they come back to the hive to tell their mates. How they do this is quite extraordinary. No other creatures apart from humans communicate so much information with so much precision, and they do it symbolically using what Karl von Frisch (1886–1982) called 'the language of bees'. Bee communication does have one essential characteristic of language, the 'sign', which comprises a 'signifier' and what it signifies, the 'signified'. Of course, there are many other features of human language that it lacks. As we will see, it only communicates one kind of signified, the location of a food source.

Western honeybees (*Apis mellifera*) live in large colonies – hives – containing between forty and eighty thousand members. Aristotle (384–322 BCE) in his *History of Animals* (*c.*350 BCE) observed that in the hive there were three 'classes' of bees – males, females and another that seemed to be neither. He was kind of right about that, but he suspected asexual generation because he and other beekeepers had never seen bees copulating. We now know that the queen copulates in flight, not in the hive.

We also now know that there is a strict division of labour: the queen who lays all the eggs, the drones who fertilize the queen, and the infertile females, workers and scouts. Workers collect pollen and nectar and return to the hive with it, but it is only in the summer that this food is available. They store food for the winter as honey in honeycombs. This means that during the summer they have to be very efficient in foraging, and not waste valuable energy flying around pointlessly. So the designated scouts go out looking for food and need to be able to communicate with the workers about its location, and to do so quickly since the food may be foraged by others, or it may go off, particularly in the tropics.

Karl von Frisch discovered how scouts communicate with workers, and got the Nobel Prize for it in 1973.[3] To communicate the location of a food source, the scout has to provide direction and distance. The nature of the food is provided by the scent that clings to the scout. Von Frisch observed that the scout communicated by means of two 'dances'. If the food was within about 100 metres of the hive, the scout produced a *round dance* (Figure 1A, overleaf), which basically said 'There's food nearby, go out and find it.' If the food was further away, and it could be many kilometres away, it was important to be as precise as possible about the location, otherwise the workers would waste a lot of valuable energy looking. In this case, the scout produces a *waggle dance* (Figure 1B, overleaf), which indicates the distance and direction with reasonable precision. The direction is specified to about 3 bits of precision (one part in eight) and the distance to 4.5 bits (one part in twenty-three). One part in eight means that north and north-east are just discriminable differences. Most navigators would not consider that to be great precision, but one part in twenty-three is precise.[4]

So how does the bee language work? What are the signifiers and what do they signify? And why does this involve mathematics? The returning scout has to calculate and dance a vector – distance and direction. Let's start with the calculation.

Distance: the scout has several sources of information which she probably combines to make a reliable estimate. First, she can assess her speed by noticing how fast she is passing landmarks, what is called 'optic flow', and there seem to be specialized neurons for this.[5] But it is slightly more complicated than this. Suppose the scout has to detour around a mountain: is the detour added to the total distance? Von Frisch thought it was. A second source of information comes from *counting* the number of

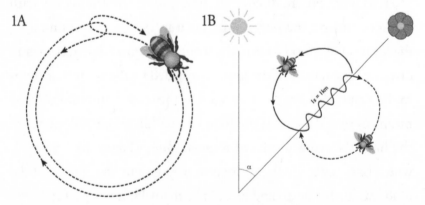

Figure 1. 1A: The 'round dance' signifying that the food is within about 100 metres of the hive. 'Just go and look.' 1B: The 'waggle dance'. The direction of the waggle portion signifies the direction in relation to the current position of the sun (α). The duration of the waggle portion indicates the distance from the hive to the food source. In the Western honeybee this is about 1 second of dance signifying 1 kilometre.

landmarks passed on the way to the food source. This method of estimating distance was originally discovered by Lars Chittka and his colleagues.

At the time, bees were thought to have cognitive abilities limited, essentially, to making associations, such as between the position of the sun, landmarks, and their internal states, such as fatigue or hunger. Chittka, then a PhD student in Germany, after perhaps too much whisky, wondered whether bees could count landmarks between the food and the hive as a way of estimating distance. This was hard to do in the wild because insect navigation couldn't be properly studied in cluttered natural environments. So with the help of fellow students and the relatively empty fields of East German collective farms, he erected large tents to serve as distinctive landmarks. By varying the number of tents as well as the distances between them, he showed it was the number of tents that shaped bee behaviour.

That bees with their tiny brains could count was such an outrageous idea, especially to Randolf Menzel, Chittka's distinguished supervisor, that it was some years before Chittka submitted a report for publication. But when it did appear, in 1995, it immediately caused a worldwide stir.

Chittka and Geiger's study[7] has been replicated in the lab. Marie Dacke and Mandyam Srinivasan from the Australian National University trained bees to forage near one of five landmarks in a 4-metre-long tunnel in the lab.[8] They also used a little methodological cunning to ensure that the bees were really using number, not some other cue. They did this by changing the appearance of the landmark. The bees trained on, say, landmark four would go to it even when its appearance was different. These findings suggest that bees were sequentially counting the landmarks in order to reach the food source on which they had been trained.

Von Frisch discovered that bees encode direction in relation to the sun's position, the solar azimuth. Notice that the bee has to take into account the solar ephemeris, the fact that the sun moves. So if, for example, the bee has to wait because it's raining outside, she has to take this into account when signalling the direction. The solar azimuth in Britain and Germany changes by fifteen degrees east to west in the course of an hour, and also the angle in the sky changes. A foraging flight may last an hour or more, as indeed may an ant's path. Thus, as Gallistel points out:

> It is essential to dead reckoning that the sun-compass be time-compensated, as von Frisch well understood. This itself implies fairly impressive calculation because to get the current bearing of the danced food location, the recruit (follower of the dance) has to add/subtract from the danced solar bearing the circular distance the sun has moved between the time the recruit observed the dance and the time at which it set out to find the food source. The time elapsed can be hours or even days. And, of course, the calculation depends on the recruit having learned the local solar ephemeris, which varies with hemisphere (northern or southern), latitude (how far north or south) and season. Again, all of this stresses the fundamental role that arithmetic plays in animal behavior. (personal communication)

Von Frisch found that bees do not actually have to see the sun. Because bee vision stretches into the ultraviolet, a small patch of blue sky enables them to assess the direction of the sun.

Now how on earth can the worker calculate where she is when she encounters a food source – the food vector? Once she has found it, let's say 10 kilometres from the hive, how does she find her way back? One way is simply to reverse the food vector, by moving so as to reduce its value until zero is reached. As I mentioned in Chapter 7 on the navigational abilities of birds, one interesting proposal for plotting these vectors has been suggested by Thomas Collett at Sussex University – a hive of bee science.[9]

The bee's brain contains a spatial array of 'accumulators' – our old friend – with different preferred heading directions, each updating according to the insect's current direction of travel. Consider the simplest case of two accumulators, one for the east–west direction, and the other in the north–south direction. During any outward path, the two accumulators sum their independent components: ten units north and three units east, for example. If the accumulators are zeroed at the start of the search, the simple sum of the contents of the two accumulators yields the total path length and the vector sum the direction – NNE.

And bees have a remarkable memory for this information. They can go back to the same food source the next day, and perhaps several days, or even months, later. How this memory is stored is still something of a mystery. Again, a likely but controversial proposal is that the bee's brain contains a map that enables it to plot the location of the food source, and remember where it is on the map and calculate the route the next day or even several days later. A map is also a way of calculating and storing the home vector.

Now the really remarkable thing is that the scout can communicate distance and direction symbolically and the workers understand what was being communicated. Figure 1B (p. 240) depicts the famous figure-of-eight waggle dance. The duration of the waggle corresponds to the distance: 1 second of waggling corresponds to 1 kilometre distance. (There are dialect differences, so that for some varieties of bee, 1 second corresponds to 750 metres.) I say duration, but perhaps the bee was actually counting the number of waggles, since there is a constant rate of waggles, about fifteen per second. So the waggle portion of the dance is the signifier for distance.

The dance for Western honeybees typically takes place on a vertical wall in the hive, and in darkness. The scout mounts a worker and buzzes loudly to attract attention and then begins her dance. The angle of the waggle portion in relation to the vertical – gravity – signifies the direction in relation to the current position of the sun. Extraordinarily, if the worker recruit is unable to go off and forage immediately, for instance if it's raining, she will recalculate the azimuth – the angle in relation to the sun – to take into account the movement of the sun across the sky. This is more complicated than it appears. Gallistel explains it like this:

> The amount by which the compass direction of the sun (its azimuth) will have changed during the rain depends strongly on the time of day at which the dance was followed and on how long the rain lasted. The sun's change in direction during any given portion of the day is given by the local solar ephemeris, which foraging bees and ants commit to memory when they first become foragers – by observations of the sun's direction at different times of day. To get the direction they must fly relative to the sun and when they finally do leave the hive, recruits must add the change in solar direction to the solar bearing communicated by the dance. This involves addition on the circle, because adding

a 360° change gets you back to where you were, not 360° further from where you were. (personal communication)

There's a lovely study from Margaret Couvillon and colleagues at Sussex University.[10] They spent two years decoding the waggle dance of thousands of honeybees in and around the campus, which is surrounded by the South Downs countryside and many city parks. They were able to map the distance and location where bees forage from month to month. The area they cover in search of food is approximately twenty-two times greater in the summer (July and August) than in spring (March) and six times greater in summer than in the autumn (October). In the summer the area they cover is 15.2 km², compared to 0.8 km² in spring and 5.1 km² in the autumn. 'There is an abundance of flowers in the spring from crocuses and dandelions to blossoming fruit trees. And in the autumn there is an abundance of flowering ivy. But it is harder for them to locate good patches of flowers in the summer because agricultural intensification means there are fewer wildflowers in the countryside for bees,' said Frances Ratnieks, professor of apiculture at Sussex University, who supervised the study. The important practical implications of the study are that 'The bees are telling us where they are foraging so we can now understand how best to help them by planting more flowers for them in the summer,' said Ratnieks.

It is worth noticing that different bee species have different foraging ranges. So the eastern honeybees (*Apis cerana*) fly up to 1 kilometre away from the nest, the dwarf honeybees (*Apis florea*) fly up to 2.5 kilometres and the giant honeybees (*Apis dorsata*) about 3 kilometres. In fact, honeybees (*Apis mellifera*) may forage up to 14 kilometres from the hive.

Ants' step-counting odometer

In terms of counting, the most surprising invertebrate is the ant. The ant brain has typically about 250,000 neurons, and weighs about 0.1 milligrams, depending on the species, with some very much smaller than that.[11] But, as we shall see, one species of ant is a champion counter.

Ants live in densely populated communities, nests; they are typically divided by caste, into queens, males and workers (sterile females), but some species have other specialists, such as soldiers or guards. Foragers go out from the nest looking for food, or building materials, and bring it back, so they need navigation skills that are equivalent to sailors' dead reckoning, termed 'path integration' in the animal navigation literature that I discussed in the chapter on birds. Like Captain Cook, the ant needs dead reckoning to determine where she is in relation to her nest, and how to get back there by the shortest and least energy-consuming route. But unlike that intrepid sailor, they do not have his instruments, a chart, a compass, a chronometer or a sextant to calculate their location, unless they have them in their tiny brains.

Dead reckoning means they have to be able to note each change of direction and how far they have gone in that direction until the next change of direction. By summing all these little paths they can calculate where they are now, say at a food source, and how they need to get back to the nest. One complication: they do not return the way they came but return by the shortest route back from the food source. We know quite a lot about how ants know direction. They can use a 'sky compass' based on the sun's azimuth, but they also need an internal clock to work out the solar ephemeris, where the sun should be at that time of day. They also have a magnetic compass based on their sensitivity to the earth's magnetic field, plus they will also learn and use landmarks near the nest. To dead reckon they have to work out the distance of each segment of the route so they can tell their current location. To find their way back to the nest by the shortest route, they will need some kind of map, or some other way of computing the shortest way home from a memory of distance and directions.

Several hypotheses have been proposed for how they estimate distance: energy expenditure or duration of the segment. Another method was proposed and tested by Henri Piéron (1881–1964) as far back as 1904. It was a very simple, but surely an implausible one, namely that foraging ants count their steps. What Piéron did was also very simple. On leaving its nest, the ant reached a point where it found food and was about to return. Piéron moved the ant a little bit further on, put it down again on the ground and observed its behaviour. The ant set off and went on a journey which was equal in length and direction to the journey it would have made if it had not been moved. In this way Piéron discovered that the ant calculated the return path to its nest on the basis of its own actions (idiothetic) and not from the surrounding landscape or local stimuli.[12]

Piéron's experiment has been replicated, and as he predicted the return track is approximately equal to the distance from the point of displacement to home.[13]

Different species have different methods. Many ant species leave a chemical trail, but *Cataglyphis fortis* live and thrive in a blisteringly hot Tunisian desert where strong winds blow constantly, so that a chemical trail would last only minutes, or even seconds. So this method of trail-making would not work. The hypothesis that the forager remembers time taken or energy expended does not work either. When the ant returns carrying a heavy load of prey or building materials, which would increase energy use and slow down the return leg, the ant still finds its way back to the nest with precision.

It turns out that Piéron was right that they have a step-counting odometer. The facts are even more extraordinary than he could have imagined.

Unbelievable? No. Matthias Wittlinger and Harald Wolf of Ulm University in Germany and Rüdiger Wehner of Zürich University carried out an ingenious experiment to test this.[14] They placed the ants in a specially created 10-metre tunnel with very few optical cues that the ants' visual system could use to estimate their own speed of travel, which in any case seems to play a minor role in route computation. They set the ants a task of finding their way to and from a food source in this tunnel. The ingenious part was to modify the ants' legs so that some ants had longer legs and some shorter legs. Ants with elongated legs take longer strides and those with shortened legs take shorter strides, so if the stride lengths are added together then those with elongated legs would go further with the same number of strides and those with shortened legs would go less far. That's a straightforward prediction that was supported by the evidence.

Now here's the really clever part of this study. It became possible to show that the ant's return path depends on the memory for the number of steps on the outward path. If the ant is allowed to go to the feeder from the (experimental) nest with unmodified legs, and then has her legs lengthened or shortened, she should miscalculate travel distance in the predicted way compared with ants with unmodified legs. And that is what happened. Those with elongated legs overestimated distance by 50 per cent (15.3 metres) and those with shortened legs underestimated the distance from the food source back to the nest by a similar proportion (5.75 metres). However, if the legs were modified before they set out from the nest, so that their calculation of distance to the nest is calculated on the basis of their modified stride length, then they should be quite accurate on their homebound journey, showing that the return path is based on the memory of the outward path.

The ants took 770 steps on their 10-metre return. If the three groups perceive these distances to be the same, then paradoxically the ants with the shortened stumps must suppose their strides to be longer than normal, while the ants on stilts must perceive their strides to be shorter! The counting mechanism, the scientists suggest, is a 'step integrator',[15] that is, an accumulator.

It is true that ants with elongating legs often walked more slowly than normal ants, but they still overestimated homing distance by about 50 per cent, suggesting that the duration of the walk was not a cue to the homeward path.

The path to food will not always be completely flat, and may involve serious ups and downs. When the *Cataglyphis* ant climbs a hill in the course of foraging, how does it take that into account when computing the home vector – the route back to the nest? Going up over a hill and down again means more steps. If the shortest route home is flatter than the foraging route, then the extra steps will mean that they overestimate the distance to the food and hence may miscalculate their location and the route home. Experiments with artificial hills show that they compute the actual map distance, somehow disregarding the ups and downs.[6,7] These ants have very clever little brains.

However, the idea that *Cataglyphis fortis* has a map in its tiny brain is not universally accepted, including by Rüdiger Wehner, the pioneering researcher who documented their step counting.

Dead reckoning itself does not require a cognitive map, but it does require that the ant converts its sense of direction in relation to the sun into something like Cartesian coordinates and displacements of them. Getting home, then, is to move so as to reduce the location vector back to zero. However, getting home by the shortest route seems to me to require a map.

Think about how Google Maps codes location data as an array of numbers, ultimately 0s and 1s, some of which refer to actual locations in the external world, such as north, south, east and west, and particular landmarks. Directions will then be numerical calculations over these data. The ant brain may be small, only 250,000 neurons, but these could code a lot of location data. Ants will also use landmarks, odours and their memory of the appearance of the nest to help guide them. Again, think about how Google Maps codes information like Street View, restaurants and gas stations; they're just more numbers.

Now *Cataglyphis fortis* could be a very special case: living in a desert, unable to use the chemical trails typical of other ant species. Would other ants count other things to help with path integration? A team led by Patrizia d'Ettorre at the Sorbonne University in Paris tested this with carpenter ants (*Camponotus aethiops*), a species found in France.[16] Could these ants count landmarks, just as bees could, to locate a food source? Figure 2 shows the set-up for counting three landmarks. The ants were placed at the start of the 'arena', a long runway. The reward was always available after the third landmark, but the position of the landmark varied on the different training trials so the ant could use distance from the start as the cue. Each ant was trained on one of five landmarks, first, second, third, fourth or fifth. Whichever landmark the ant was trained on, it would get faster at locating the food source on subsequent trials. On the test trial, the arrangement of landmarks was different from any of the training trials. Nevertheless, when the ant was trained, for example on the third landmark, it would go to the third landmark on the test trial even though the spatial arrangement of landmarks had changed. See Figure 2.

A more direct test of ants' number sense is to use our familiar match-to-sample paradigm (see Chapter 1). With ants you sometimes test colonies rather than individuals. In the following study, colonies of 200 to 2000 members of the species *Myrmica sabuleti* were collected from abandoned quarries in Belgium and divided into twelve experimental colonies for an

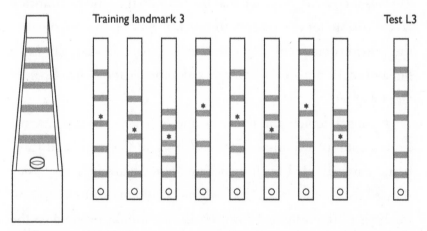

Figure 2. Training an ant to find food at the third landmark (L3). Each training trial rewarded (*) the ant at L3, but the distance from the start (o) of a runway varied at each training trial so the ant could not use distance as the cue to the reward. The test used a new arrangement of landmarks and no food reward, and the ant reliably searched at the third landmark.[16]

experiment by Marie-Claire Cammaerts from the University of Brussels and Roger Cammaerts of the Natural and Agricultural Environmental Studies Department (DEMNA) of the Walloon Region in Belgium.

The ants were trained by rewarding them near a display of one, two or three paper squares, but far from similar displays of one more square, in a foraging tray for seven, twenty-four, thirty-one and forty-eight training hours. Then the ants were tested in a separated tray with paper stimuli that differed in shape (discs instead of squares), colour, size or arrangement. Ants learned the correct numerosity after seven hours of training, but got even better after more training.

Counting in the lab

There is increasing evidence that bees can do the kind of counting tasks possible for creatures with much bigger brains described in the previous chapters. For example, bees can successfully carry out a numerical match-to-sample task pioneered by Otto Koehler (see Chapter 1).

Figure 3 depicts the result of one of a series of experiments from a team of scientists from Würzburg in Germany and Canberra, Australia, led by Jürgen Tautz and Shaowu Zhang. They showed that honeybees (*Apis mellifera*) could match to sample on the basis of the number of objects in the sample, even when the colour, object type and arrangement in the sample were different from the choices offered, up to five objects.[18]

With small numerosities, bees can also add and subtract. Bees are sensitive to colour (including, as I said earlier, to ultraviolet). In the following study, colour instructed the bees to add or to

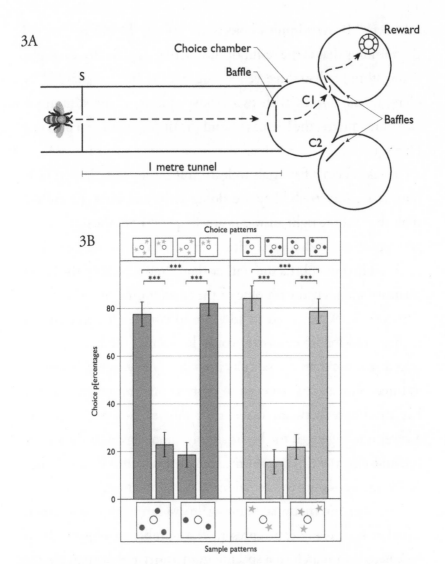

Figure 3. 3A: A typical set-up for testing bees' match-to-sample ability. The bee subject sees a sample, S, and can then select a match from two choices, C1 or C2. If correct, she will get a reward. 3B: Results. *** denotes a statistically significant difference at p<0.001 between columns. This means that the result would occur only one time in a thousand by chance. Notice that the sample is always different in object type and arrangement from the choices (in the study, colour was also different).[18]

subtract.[19] If the stimulus objects are yellow, the bee has to find the stimulus that is the result of subtracting one. So, for example, if the stimulus has three yellow squares the bee is rewarded by going to the display with two yellow squares. If the stimulus is two blue squares, the bee has to find the display with two plus one blue squares.

It takes bees a few trials to learn this rather complicated task, but after thirty trials they are doing well, and after a hundred trials they get the right answer nearly 80 per cent of the time. There was a nice methodological trick to make sure that the bees were really adding or subtracting rather than simply picking the larger quantity when cued with blue or the smaller quantity when cued with yellow. When cued to add one to two, the correct answer is three and incorrect answer could be four. If the bee is simply selecting more than the sample, it could equally well choose four as three. Similarly, when cued to subtract one from three, the bee has to choose between the correct answer two and incorrect answer one. Again, if the bee is simply choosing fewer than three it could just as well choose one. But the bees don't do this. They really do choose the correct addition or subtraction.[19]

Another international team again using match-to-sample, is whether bees really can represent the number of objects. Their task here is to match to a specific numerosity rather than choose the larger or the smaller (see Figure 4).

One of the most interesting characteristics of human counting is the ability to conceive of zero. We know that a symbol for zero was not known by the great Greek, Babylonian or Egyptian mathematicians and was not introduced into Europe until the twelfth century following an invention in the Indus Valley in the seventh century. This suggests that the concept of zero is

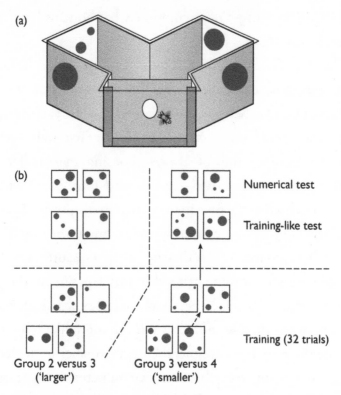

Figure 4. The training phase consisted of thirty-two trials during which stimuli were varied from trial to trial, keeping the rewarded numerosity constant. The groups 2 vs 3 (larger) and 3 vs 4 (smaller) were always rewarded for choosing three. After training, bees were subjected to two non-reinforced tests. In a training-like test, they had to choose between the numerosities experienced during training displayed through novel stimuli. In a numerical test, they were confronted with a novel situation in which the rewarded numerosity (three) was opposed to a novel one (four for the 'larger' group, and two for the 'smaller' group).[20]

surprisingly difficult to master. But as we have seen in Chapter 4, chimpanzees have been shown to possess the concept and can link it to a symbol. It has rarely been observed in other creatures, just monkeys and some birds. I suspect this is because no one has thought it worth testing in these small brains. But what about bees? Can their tiny brains compute what humans have taken so long to symbolize?

With sufficient methodological cunning, it has been shown that bees do have a sense of zero. The way to demonstrate this is to think of zero as being at the lowest end quantity of the sequence of positive whole numbers (see Chapter 1). First, train bees to follow the constant rule of always choosing the smaller of two numerosities, let's say between one and four. Now, when they were presented with one item versus an empty background, a situation they never saw during the training, they preferred the empty background, which they treated as a quantity smaller than sets of one, two, or more items.[21] In this experiment, there is also an interesting echo of Weber's Law, namely the bigger the difference between the two numerosities the more likely the bee will choose the zero. For example, their performance improved as the magnitude of difference between two numerosities increased (0 vs 6 was easier than 0 vs 1). Weber's Law implies that indeed the bees were treating the numerosities as a sequence of magnitudes, just as we do.

Since many flowering plants which depend on insect pollination have a constant number of petals, pollinators such as bees evolved a certain ability to discriminate flowers based on the number of petals, to about four; for larger numerosities, bees probably use the overall shape. Perhaps this is an example of co-evolution – flowers evolve to attract bees by having the petal arrangements that attract bees, symmetrical and relatively few petals – and bees evolve to be able to identify flowers that are a rich source of food.

Evidence from route finding shows that bees understand 'larger than' and 'smaller than' rather concretely in terms of distance. One notable experiment suggests that bees possess a cognitive framework for interpreting various dimensions in terms of *more than* and *less than* abstractly. Maria Bortot, Gionata Stancher and Giorgio Vallortigara at the University of Trento in Italy showed that bees trained to select from two panels the one with larger numerosity would then spontaneously transfer to selecting one of two panels with the same number of dots but different total dot area.[22] This is impressive, but nevertheless, as we have seen earlier (see especially Chapter 1), when numerosity changes, many other dimensions change as well. For example, maintaining the same total area means increasing the total length of edges. It seems to me possible that bees could be tracking one of the physical dimensions from the numerosity training to the size test.

These studies show that bees can *learn* to use numerosity and computations over numerosities in lab tasks. These are not the kind of problems that bees in the wild confront. The computations that bees do in the course of their daily lives, calculating route vectors and quantities of food at specific locations, are far more complex.

Beetles counting their rivals

Here's a quite different kind of study where the subject has to count a succession of biologically relevant objects. Like other insect species, male mealworm beetles (*Tenebrio molitor*) adjust their reproductive behaviour according to the perceived risk of 'sperm competition' – that is, they will increase guarding their mate when there are more rival males around. The question here is, do they actually count the rivals?

Pau Carazo, Reyes Fernández-Perea and Enrique Font from

the University of Valencia in Spain tested this idea with a very ingenious experiment. They staged matings between virgin males and virgin females, and presented rival males to the male subject sequentially – i.e. one at a time – but with this clever twist to see whether it was, let's say, the quantity of maleness or the number of males that was critical. In the one-rival condition, the subject would be exposed to the same rival four times, three minutes a time, with a two-minute interval between exposures, while in the four-rival condition, the subject would be exposed to four different males on the same time schedule. The result was clear: the male spent more time guarding the female after mating when exposed to four rivals than one, though guarding duration didn't increase linearly with one, two, three and four males.[23]

Cicadas counting prime numbers

As I mentioned in Chapter 1, in the 1985 sci-fi novel *Contact*, the American scientist Carl Sagan imagined the first contact between humans and an extra-terrestrial civilization. Ellie, the protagonist, knew that the signal must be a message, because it contained the sequence of prime numbers and so must have originated from an intelligence. Only an advanced civilization could know about prime numbers, she reasoned. Before Euclid in about 300 BCE, there is no record of people knowing about or being interested in the sequence of prime numbers. Euclid showed that there are infinitely many primes (*Elements*, Book IX, Proposition 20), and that every integer can be written as a product of primes in an essentially unique way – nowadays called 'the Fundamental Theorem of Arithmetic'. Many of the greatest mathematicians – Euler, Erdös, Fermat, Gauss, Hardy, Mersenne, Ramanujan, Riemann – have devoted much time and thought to understanding primes, building upon what had previously been discovered. It therefore seems improbable, to say the least, for other creatures, especially those with very small brains, not to mention with no history of mathematics to draw upon, to have a sense of prime numbers. But wait.

Let me start with a personal story.

Many years ago, I was walking with my family in a remote part of Wilsons Promontory, a lovely national park not far from Melbourne, Australia, and we were deafened by cicadas. I now know that each insect can produce a noise of 120 decibels – about the sound of a heavy metal band – and here there were millions. At the time, all I knew about cicadas is that children at my daughter's school gathered them for show and tell, and collected the larval cases from which they emerge and painted them in gaudy colours. What I didn't know was that this deafening experience only happens once every seven years, which is why the children were excited. It turns out that there are many species of cicada that emerge from their underground nymph state in a prime number cycle – yes, a prime number cycle. In the US, for example, there are periodical species that emerge after thirteen and seventeen years (making them, incidentally, the longest-lived insects). So why a prime number periodicity? And how do cicadas calculate it?

The why question is actually easier to answer. There are evolutionary reasons, and we know, famously, that evolution is interested in three things: food, sex and death. The periodicity is about all three, but we need to understand the lifecycle of a cicada, like the Australian ones we encountered, the 'green grocer' (*Cyclochila australasiae*).

The cicada eggs become nymphs – rather similar in shape to

the adult form, but lacking wings. The nymph will spend seven years underground drinking sap from plant roots before emerging from the earth as an adult, called the 'imago'. Here they have food and are relatively safe from predators.

The adults live for about six weeks, fly around, mate, and breed over the summer. It is vital that they emerge at the same time so that there are mates around. Only the male makes his deafening song, which females find attractive. If the female is attracted, she will make a clicking sound that will attract a male. Mating will occur, and the fertilized eggs are deposited on a tree in whose roots the nymph will make its home for the next seven years.

Emerging en masse also increases an individual's chance of survival. Predators cannot possibly devour the lot of them when thousands or millions emerge at the same time, and the larger the brood, the more likely is 'predator satiation', so both brood size and brood synchronicity are adaptive in avoiding death.

But why did prime number cycles evolve? An answer was suggested by Stephen Jay Gould: the key factor is that seven, thirteen and seventeen cannot be divided evenly by any smaller numbers (except one). Prime cycles, Gould notes, have a major evolutionary advantage over cycles that are multiples of smaller numbers of years, and for a simple reason: they make cicadas more elusive. 'Many potential predators have 2–5-year lifecycles . . . Consider a predator with a lifecycle of five years: if cicadas emerged every 15 years, each bloom would be hit by the predator. By cycling at a large prime number, cicadas minimize the number of coincidences (every 5×17, or 85 years, in this case).'[24] Similarly, with the green grocer's seven-year cycle, the five-year predator would only coincide every 5×7 years, and a two-year predator every 2×7 years.

So, how on earth do cicadas keep track of their prime number cycle? Bear in mind that most insect cycles are within a single year and depend on the daylight and temperature. Since the nymph is underground these indicators scarcely apply. Note also that any biological clock must recognize a cycle and be able to count the number of cycles, and the clock counter must connect with the neuroendocrine pathways that control the response, here the emergence of the nymph and its transformation into the imago.

Richard Karban at the University of California, Davis, and his colleagues asked what controlled the seventeen-year lifecycle of the *Magicicada septendecim*, a cicada found in North America. It was unlikely to be changes in daylight or temperature that are hard to detect underground. They wondered whether instead of the direct influence of these factors, the cicada was indirectly affected through the seasonal changes of the availability or quality of the sap that sustained the nymph underground.

What they did was really quite remarkable. They transported in potatoes fifteen-year-old nymphs to the roots of 'a judiciously selected' peach cultivar that was capable of double cropping under appropriate conditions – that is, they could speed up the annual cycle of the peach to see if this caused the nymph to emerge early. If it did, then this would show that annual changes are governed by the sap that sustains the nymphs over the seventeen (*septendecim*) years. That's what indeed they found in the nymphs feeding on the speeded trees. Karban and colleagues conclude that 'cicadas do not time their 17-year pre-adult development through endogenous time keeping . . . but rather, through counting the number of host phenological and seasonal cycles'.[25] At the end of the study, Karban writes that 'he has always dreamed about tricking them [cicadas] into emerging early for most of his adult life'.

So we now know what they count, but we still don't know how they count. One possibility that appeals to me is, of course, the accumulator system. It would require that the selector opens the gate to increment the accumulator in some way synchronized with seasonal phenological cycle, that is, once a year until the accumulator level has reached the species-appropriate prime number, seven in the case of the green grocer, seventeen in the case of the *Magicicada septendecim*.

Spiders

The spider's brain is tiny, smaller than a honeybee's at about six hundred thousand neurons, but it too has a complicated behavioural repertoire – notably, of course, spinning webs. Some stalk their prey, rather as cats do, such as the jumping spider (*Portia fimbriata*). It watches the motionless prey from a distance, before engaging in a circuitous path, which often involves losing sight of it, so as to appear near the top of the orb-web. From this position, the jumping spider hangs down from a line until she is able to grab the prey in the centre of the web.[26] And adult orb-weaving spiders vary in body mass by four hundred thousand times (including those near the lower size limit for spiders in general). For example, a nymph of the orb-weaving spider, *Anapisona simoni*, with a body mass of less than 0.005 milligrams, appears as a speck of dust to the unaided eye, and yet there is no evidence of inferior performance due to its very small size.[27]

One thing that web-weaving spiders might count is the number of prey trapped in their web. This can be tested by removing prey and seeing if the spider goes searching for it and whether search time depends on the number of prey removed. This is exactly what Rafael Rodríguez and his colleagues from the University of Costa Rica investigated.

They studied the large golden orb-web spiders (*Nephila clavipes*), so called because their silk looks golden in sunlight. *Nephila* accumulate prey 'larders'. In the wild, the spiders that lose more prey items search for longer intervals, 'indicating that the spiders form memories of the size of the prey larders they have accumulated, and use those memories to regulate recovery efforts when the larders are pilfered . . . *N. clavipes* mainly lose prey to kleptoparasitic *Argyrodes* spiders.'[28] However, search time might depend on the total amount of prey 'stuff' (mass), not on the number of individual prey items. This is a methodological problem we have repeatedly encountered in previous chapters: are animals tracking quantity or number? Hank Davis and Rachelle Pérusse, in a major review on numerical cognition in animals, proposed that animals only use numerosity as a 'last resort' strategy.[29] They didn't consider spiders in their review, but is it true of these spiders?

Rodríguez and his colleagues allowed the spiders to accumulate and store prey – in this case mealworm larvae – on their web by dropping a larva on the sticky spiral of the spider's web so that the spider could perform its normal prey capture behavior with each prey item – locate prey, extract it from sticky spiral, bring it to the hub, wrap it in silk, and secure it to the hub and settle into feeding for thirty seconds.

In the numerosity test, spiders were allowed to accumulate larders of one, two or four small prey, and then the entire larder was removed and search times recorded. In the spider stuff test, the spiders were allowed to form larders of one single prey item of either small, intermediate (= two small items in mass) or large size (= four small items in mass), and then the prey were removed.

It turns out that search time is highly dependent on the number of items, but less so on the mass of the prey. The spider, understandably, is interested in both the mass and the number of prey in its larder, but is more interested in the number.

Rodríguez and colleagues speculate that the *Nephila* may be using an accumulator mechanism to count the number of prey one by one, since each prey item was attached to the hub with a single silk line and the spiders fed on only one item at a time.

Back to jumping spiders and their legendary documenter, Robert Ray Jackson of the University of Canterbury in New Zealand. In a special issue of *Natural History* magazine devoted to research on spiders, Robert was asked to write the introduction. He wrote:

> The salticids [jumping spiders] are my favourite spiders because in their behaviour, they defy the conventional wisdom that spiders are little more than instinct-driven automatons. The ultimate salticid is *Portia*, a genus of tropical jack-of-all-trades spider. To catch its prey, *Portia* can build webs, but it also hunts without them. Often the spider will even enter the web of another spider and capture it. These extraordinary raids demonstrate the spider's problem-solving skills. For example, there are some webs that *Portia* prefers to enter from the top. If it spots such a web from below, a passing *Portia* looks around, then takes an indirect route through the vegetation to reach a point above the web – even when this means initially going away from the prey and venturing where the prey is temporarily out of view. Once in the web, *Portia* uses vibratory signals to deceive and manipulate the resident spider, often varying and combining these

signals by trial and error to slowly coax the prey closer before attacking.

Portia must be [the resident] spider's worst nightmare. It hunts with mammal-like cunning, has excellent eyesight and a predilection for the flesh of spiders, including jumping spiders.[30]

Portia africana, as the name suggests, is an African spider that preys on other spiders. As Jackson noted, *Portia* has very good eyesight and after seeing a scene with prey, it may initiate a detour during which the prey is no longer visible and then attacking. To do this successfully *Portia* has to remember what and where the prey is. But does *Portia* remember the number of prey? This is the question Jackson and his colleague Fiona Cross asked in a very ingenious experiment.[31]

They built an apparatus with two towers. First was a starting tower into which *Portia* was placed. From there it could see a scene with some prey in it. The spider could then do what comes naturally, detour to another location from which to attack the prey. The apparatus was designed so that the spider could only end up at the top of the viewing tower, where it could see the prey again and could attack. However, and here's the clever part, either the same scene was displayed or a different scene. If the scene is different, will the spider notice this, and will this affect its behaviour, for example by delaying the attack?

Using this methodology enabled the experimenters to manipulate the properties of interest, and noticing a difference between the remembered scene and the new scene is measured by the latency to attack. So, if the number of prey in the scene is kept the same but the size of the prey doubled, will the subject spider delay attack? Answer: no. If number of prey is the same but their arrangement is changed, will the subject spider delay attack? Answer: no. If the number of prey in the scene is changed? The answer is yes, if one prey is changed to two, or two to one; and if one is changed to three or three to one; one changed to four or four to one; similarly two to three, two to four, two to six, but not for three to four or three to six. So for small numbers, and if the ratio difference is sufficient, then the spider remembers the exact number of prey and can compare the memory with the new scene. This suggests that *Portia*, like the rest of us, has a visual enumeration system for small numbers – in this case it seems to have a limit of three – and a larger number system for four and above, but this is only effective when the ratio difference between the two numerosities is large enough. That is, for these larger numbers, Weber's Law applies.

In one species, *Portia africana*, the tiny juveniles (2.5 millimetres in length) hunt another species of spider but they practice 'communal predation'. The juvenile will go looking for the nest of a prey and will note how many others have settled on the nest. It turns out that they prefer nests where there is one other *Portia*, not zero, not two or three.[32] Ximena Nelson and Robert Jackson from the University of Canterbury in New Zealand, who carried out this fascinating study, believe that 'Number appears to be a salient cue when *Portia* makes settling decisions,' but they go on to say that:

> there is no basis for proposing that *Portia*'s decisions are based on anything especially close to true counting [because] true counting is based on representing numbers as properties of sets that stay the same even when the identity of the

particular objects in a set change, but the expression of numerical competence we have investigated seems to be linked tightly to objects with highly specific characteristics (i.e. other *Portia* already settled at [a prey] nest).[32]

As I have argued in previous chapters, the difference among species is not so much in their ability to count but in the range of things that they can count, and also how high they can count. So, from this point of view these juvenile spiders are counting, admittedly not very high, but they can only count other *Portia* or prey packages in other spiders' webs. No one has tested whether the *Portia* represented two other *Portia* and two prey in the same way. But it is worth remembering that in the wild they would be counting different *Portia* on each occasion, and different prey packages in different configurations on different nests. Of course, it is possible that they can count other things, but no one has tested this yet.

Insect brains

Insect brains are very, very small, as I have stressed. The question is, can brains with 1 million or fewer neurons actually count? Can they implement the accumulator system? Vera Vasas and Lars Chittka have modelled a tiny bee brain with just four 'neurons' – simple mechanisms that do just one job.[33]

Now the bee counts sequentially. She crawls over a stimulus, for example, the petals of a flower to establish the numerosity of the set of petal.[34] The computer model of bee counting by Vasas and Chittka starts with *brightness 'neuron'* which is the equivalent of the selector: it simply notes the change of brightness as the bee traces a path across the display, which is the method bees use in the lab experiments described earlier. When it goes from dark to light it is this change that is counted in an accumulator 'neuron', which accumulates information about changes in brightness over a longer period, and the 'evaluation neuron' establishes the numerosity of the count. Only four virtual neurons are needed to model bee counting.[33]

Cuttlefish

The cuttlefish, a cephalopod, has a very big brain, but we don't know exactly how big. It is very difficult to get a sensible estimate because its nervous system stretches into its tentacles, which can operate independently of a central nervous system, and as with the octopus, another cephalopod, it is not clear where the brain itself begins and ends. Indeed, in the octopus the majority of neurons are in the arms themselves – nearly twice as many in total as in the central 'brain'. According to the Israeli scientist Binyamin Hochner,

> The size of the modern cephalopod nervous system normalized to body weight lies within the same range as vertebrate nervous systems – smaller than those of birds and mammals, but larger than those of fish and reptiles. Comparing the total number of neurons, a variable that may be more relevant to neural processing, the octopus nervous system contains about 500 million nerve cells, more than four orders of magnitude greater than in other

molluscs (garden snails, for example, have around 10,000 neurons).[35]

There have been studies investigating the cognitive capacities of cephalopods from the time of Aristotle, who wrote in his *History of Animals* that 'The octopus is a stupid creature, for it will approach a man's hand if it be lowered in the water' and so gets caught and eaten. By contrast, he thought that cuttlefish (*Sepia*) were smart. 'Of molluscs the sepia is the most cunning, and is the only species that employs its dark liquid for the sake of concealment as well as from fear: the octopus and calamary make the discharge solely from fear.'

More recent research, however, suggests that actually cephalopods, including octopi, are very smart, can learn many tasks, and may even have consciousness. As far as I know at the time of writing, there has been only one study of cephalopod numerical ability, and that is in the cuttlefish. This is by two Taiwanese scientists, Chuan-Chin Chiao and his colleague Tsang-I Yang.[36]

They asked whether juvenile cuttlefish (*Sepia pharaonis*) would show a preference for a larger quantity when faced with two-alternative forced choice tasks with live shrimp (1 vs 2, 2 vs 3, 3 vs 4 and 4 vs 5). Cuttlefish actively prey on shrimp, and when they see them they shoot out two tentacles to capture them. The cuttlefish made all the discriminations, including 4 vs 5, which is beyond the limit of many other species, including some monkeys (see Chapter 4). It's worth noting that as the ratio difference between the two numerosities decreased, reaction times *appeared* to increase: Weber's Law again, though it has to be said that there is no statistical test in the paper that shows a systematic effect.

Yang and Chiao conclude that 'Our finding that cuttlefish suc-ceeded when the tests were 1 versus 5 and 4 versus 5 implies that cuttlefish are at least equivalent to infants and primates in terms of number sense, and their number discrimination is likely to be an analogue magnitude mechanism of numerical representation and perhaps a continuity of the number system.'[36]

However, there is one very interesting fact about these cuttlefish, and that is that their numerical choice depends on whether they are hungry or not. When the choice was between one large live shrimp and two small shrimp, they chose one large shrimp when they were hungry and two small shrimp when they were satiated.

Cuttlefish certainly seem not to behave like the guppies described in Chapter 8 on fish. Guppies don't show a ratio effect when the numbers to be compared are four or below. It is only the numbers larger than four that show this effect, and that has been an argument for proposing at least two systems of numerical esti-mation, the small number ('object file') system and the larger number ('analogue magnitude') system. As I mentioned in Chap-ters 1 and 2, when the animal's Weber fraction is 0.25 or smaller, then all comparisons of numerosities in the range 1 to 4 (e.g. 2 vs 4 is 0.5 and even 3 vs 4 has a Weber fraction of just 0.25) won't show a ratio effect, since they will all be equally easy. This means that a special mechanism for low numbers is not necessary, at least for humans (see Chapter 2). Now we don't really know what the Weber fraction is for cuttlefish. This is just one study of cuttlefish numerical ability, and only one kind of test. We don't know, for example, how cuttlefish would manage with a match-to-sample task with all the appropriate controls, or with objects that are not food items. Would we still see this continuity, and would there be a limit at 4 vs 5? It would be very interesting to know.

Mantis shrimp

The first thing to say about the mantis shrimp (*Neogonodactylus oerstedii*) is that it is not really a shrimp. Unlike the common shrimp, it has extraordinarily complex eyes at the end of stalks. They have powerful raptorials (forelegs adapted for grasping prey, hence 'mantis') that are used to attack and kill prey by spearing, stunning or dismembering. Some mantis shrimp species have specialized calcified 'clubs' that can strike with great power. They are general poorly understood, but one key part of their mysterious lives has recently been revealed by Rickesh Patel and Thomas Cronin of the University of Maryland.[37]

The mantis shrimp leaves its burrow to forage and find mates, returning quickly to avoid predation. These trips away may be as much as 4 metres, which is actually a substantial distance for animals typically around 3–5 centimetres long, though not a lot compared with ants or bees. So how does it plot its path to food, and find its way home? It turns out that Patel and Cronin have shown that these little invertebrates use path integration, just like the birds we saw in Chapter 7 and the bees and ants we have seen in this chapter.

The demonstration is beautifully elegant. The shrimp is placed in an arena with a fictive burrow so that its visual field can be manipulated. So, for example, a board is used to conceal the sun or a mirror to displace the position of the sun. With the sun concealed, the homeward route is still correct, showing that the shrimp can use other cues. However, when the sun's position is reversed by mirroring, the shrimp goes in the opposite direction following the sun. A similar result was obtained by rotating the polarization pattern in the artificial 'sky'. First, they try to use the sun. If there is no sun, they use the polarization pattern, and if that isn't possible, they use whatever clues they have, including internally generated ones. When none of that works, they use landmarks which they can see with their extraordinary eyes.[37]

The organization of the mantis shrimp brain is remarkably similar to those of insects, and the way it calculates the route home may be very similar.

Snail counting

To test the counting ability of snails takes great ingenuity and great patience, which two scientists at the University of Padua demonstrated in abundance when they tested the Mediterranean dune snail, *Theba pisana*, which:

> occupies a harsh habitat characterised by sparse vegetation and diurnal soil temperatures well above the thermal tolerance of this species. On sunny days, the sand can reach temperatures of 75°C, way outside its thermal limit, so to survive, a snail must locate and climb one of the rare tall herbs each dawn and spend the daytime hours in an elevated refuge position.[38]

Bisazza and Gatto guessed that the snail would prefer a larger choice of refuges and set out to test their ability to distinguish between options.

In their experimental set-up, the refuges are represented by vertical lines variously spaced. The choices are levels of difficulty from very easy (4 vs 1) to very hard (5 vs 4) to very, very hard, which most of the animals in the book cannot manage (6 vs 5). To check whether the snail was using the total area of the lines, a second experiment tested its ability to choose between black squares of the same ratios. In this latter case the snail could only manage the easiest ratio. Various other controls were explored in separate experiments. Bisazza and Gatto conclude that only primates and a few other vertebrates appear *more* accurate in discriminating discrete quantities (chimpanzees, rhesus monkeys and pigeons), while many other species show much lower numerical acuity (e.g. red-backed salamander, 2 vs 3; horses, 2 vs 3; and angelfish, 2 vs 3). So this snail is a kind of numerical champion despite having a very tiny brain indeed, perhaps only ten to twenty thousand neurons, the fewest of the animals reported in this book.

So can invertebrates count?

The one with by far the largest brain of those I have described, the cuttlefish, has been studied least. From what we know so far, it is able to distinguish numerosities of shrimp up to five. Nothing is known about its abilities beyond five, and there have been no match-to-sample tests but only the test of comparing numerosities. And it hasn't been established that the result of counting, if that is really what it is, can enter into computations.

Spiders are also little tested, but they can remember the number of prey in their web up to at least four, and this seems to be an ability used in the wild when their 'larder' is raided by other kleptoparasitic spiders. The application of spider counts is very limited in both what they can count and are interested in counting. Can spiders compute with their counts? We don't yet know.

There are relatively few studies of the numerical abilities of ants. They seem to be able to match to sample the number of abstract shapes, and to add one to a number. How this ability is used in the wild is not known. What is known is that a species of desert ants count their steps, thousands of them, to calculate the distance between the nest and a food source. So the results of ant counting can and do enter into computations.

Bees are perhaps the most interesting because they alone in the animal kingdom, apart from us, communicate numerical information *symbolically* by means of their waggle dance, indicating distance from hive to a food source. To do this they must have calculated the distance, and one way of doing this is by counting landmarks. The mere achievement of plotting their route from hive to food and back again requires computations that must at bottom be numerical. Again my model here is Google Maps, as I argued for bird navigation. In it, a map is stored as an array of numbers and directions are the result of calculation over those numbers. This will provide a current location, and enable the bee to calculate the 'beeline' (home vector) back to the hive. These calculations will come with errors of direction and distance, but will normally be good enough to get the bee close to home. No doubt visible landmarks, and perhaps the smell of the nest, may then be needed to accurately locate the hive. A Google map can be requested to include landmark information, such as restaurants and gas stations, as well as street views, and is exceptionally detailed. The question seems to me not to be whether bees have a map in their brain, but how detailed and complete it is. Does it just store key vectors for returning to a food source (restaurant), a few key landmarks such as the entrance to the nest (selected street view), and a remembered calculation for returning to the nest by the shortest path?

Bees on test are also able to match to sample on small numerosities and are able to learn to add or subtract one when presented with a cue to do so.

The simple accumulator mechanism described in Chapter 1 could accomplish these tasks. Recall that it contains three components: the accumulator itself, a memory for each accumulator to store the final count, and a selector to identify the things or events to be counted. Accumulators can store inputs from continuous quantities but the outputs into memory must be in a 'common currency' with counts of discrete objects or events in order to calculate frequencies and probabilities. The main difference among species is, I have argued, about what can be counted, not about the counting mechanism itself.

The mantis shrimp also uses external cues such as the sun's azimuth or, on cloudy days, the pattern of polarized light from the sun, to provide direction, and enables it to calculate the beeline home.

What can be counted will depend on what is important to each species, though it is likely to be one or more of food, sex and death. When an organism is hungry, food will be selected; in the breeding season, sex; and in conflicts with rivals or predators, avoidance of death may trump other considerations. In the case of the dune snail, for example, keeping off the baking hot sand is the priority.

For bees and ants when foraging, it is the distance and direction of food (workers are in both cases infertile females, so sex is not relevant, and workers seem willing to sacrifice themselves for the good of the colony, so avoiding death may not be an individual's priority). For the male beetle, the accumulator counts male rivals, that is, it's sex and reproduction for him. For cicadas the accumulator counts seasonal changes in tree sap, and it's about sex – males and females must emerge from underground at the same time in order to mate. But it's also about death: by using prime number cycles they have a better chance of avoiding cyclic predators. We don't know about counting, sex and death with cuttlefish, only about food. How it is implemented in the brains of bees and other invertebrates is still to be discovered.

So, are big brains better? Not for counting, as we have seen. The accumulator is a tiny system that can fit into the smallest brain so even ants and snails can count. Bigger brains enable more different types of thing to be counted, which is the function of the selector.

CHAPTER 10

WHAT COUNTS?

Most of us, when we think of counting, think of a process that is intentional, purposeful, conscious and usually accompanied by counting words. This definition rules out most if not all animal studies. They don't have counting words, apart from Alex the Parrot (Chapter 6). Without a verbal report it is hard to tell whether a numerical assessment is intentional or just something that is automatic, like seeing the world in colour. Purposefulness could be another conundrum. Shoaling in small fish or retreating when outnumbered are certainly adaptive, but are the fish behaving purposefully? Whether non-human animals are conscious is an even more difficult conundrum. We may be happy to ascribe consciousness to primates, pets and perhaps birds, but less happy to ascribe it to insects or fish.

So, let us forget how we verbal humans count, and focus instead on how other creatures extract numerical information from their environment, to 'understand the language of the universe', or at least a small, but fundamental, dialect of it concerned with sets and their sizes. This, as I have argued, is adaptive for food, reproduction, competition and navigation.

Our investigation of the evolution of counting has been governed by two guiding principles stated in Chapter 1. Here I followed Gallistel in his proposal for assessing whether an animal, or a human, is actually capable of representing number in its brain.[1] He set out two criteria, one of which is: can the animal represent numerosity as a distinct property of a set, separable from the properties of the items that compose the set?

For him and me, and innumerable philosophers, numerosity is an abstract property of a set. The philosopher Marcus Giaquinto identifies numerosity explicitly as the size of a set, so the sets of three apples, three chimes or three kisses have the same set size even though the items have nothing in common.[2] As we have seen, experimenters, following the pioneer Otto Koehler, have tried to ensure that an animal's task cannot be solved by non-numerical means using other visible properties of the items composing the set, such as the volume or area covered by each item, or, and this is important for many animals, their odour, even though in nature numerosity and the other dimensions typically co-occur. More food items means more food volume and sometimes more food smell. Some experiments require the animal to keep track of the items presented sequentially – for example, dropped into a bucket or disappearing behind a screen – so that it is necessary to keep in memory the number of objects. In particularly clever experiments, a set presented as a sequence of objects has to be matched with the numerosity of a display with all the objects presented at once, as Koehler did with his birds.[3]

Although we humans can count pretty well anything, this does not mean that non-human animals can count any objects or events, or actions, or match across the modalities of vision, audition or action. They tend to spontaneously count only those items that are relevant for their survival and reproduction. This means we should not rule out animal counting simply because they can only count one or a few types of objects. They are still abstracting the particularities of the objects they can count. A bee counting petals is abstracting the particular flowers in their particular location and background. Plus the animals I describe *can* learn to count unnatural objects – arrays of dots, for example – or actions they would never have to do in the wild, such as lever pressing; but they usually do so with difficulty and after long training.

In the wild, it is not enough to be able to count the items in the set; the animal has to be able to do something useful with the result of the count. They have to be able to calculate; that is, to do what Gallistel calls 'combinatorial operations' of a certain type, 'isomorphic to the arithmetic operations that define the number system (=, <, +, −, x,÷)'.

Most animal studies, as we have seen, have set out to demonstrate that the animal is able to represent set size, in most cases implicitly using three arithmetical operations. Animals are set the task of deciding which of two sets is numerically larger, or more rarely, numerically smaller. The operations implicitly invoked are < or >. In the other paradigm, match-to-sample, the animal is tasked with finding the set that was equal to the sample, that is, =. In the lab, it was rare for the other arithmetical operations tested.

What other animals have that we also have

I have argued in this book that we humans and other creatures studied have in their brains an accumulator mechanism. This is a small mechanism that doesn't require many brain cells, as Rose showed in the case of the túngara frogs (see p. 202), and requires four elements to model, as Chittka has demonstrated (see p. 267). Because it's small and efficient, and all built to the same design, brains can afford more than one of them. Bee brains, for example, may have one accumulator for landmarks to help them estimate distance from the hive to a food source, and another accumulator for counting petals on flowers so that they can go to the most likely food source.

The originators of the accumulator proposal, Warren Meck, Russell Church and John Gibbon, suggested that the accumulator could operate in two 'modes', one to count objects and events, the other to measure duration. If there is duplication of accumulators, one could be dedicated to numerosity and another dedicated to measuring duration. In this case, the two accumulators would need to communicate in a common language to calculate rates or probabilities (duration/numerosity).

Multiple accumulators limit but do not completely eliminate the essential abstractness of the animal's conception of number. In the case of the honeybees, each set of landmarks, for instance, could contain different objects; each flower will be unique in some way, including location and background.

The component requiring more brain cells and more experience in the accumulator system is a selector that determines the objects or events to be counted. In Chapter 1 I envisaged the task of comparing the number of sheep with the number of goats. This is quite a difficult problem – at least for me – because both are four-legged hairy grass eaters about the same size, sometimes sharing a field, and there are many species and breeds of each. The selector needs to know quite a lot about sheep and goats to assign each object to their respective sets, and this knowledge takes up a lot of brain.

For full human-style counting the selector must first identify the apples, the chimes and the kisses, then ensure that three apples, three chimes or three kisses are routed into a single accumulator or, perhaps, coordinated accumulators that, minimally, enable the numerosities of apples, chimes and kisses to be evaluated and compared.

An accumulator mechanism in the brain will respond more with more objects or events experienced. Some evidence for such a mechanism has been observed in human,[4] monkey[5] and cat[6] parietal cortices, as we described earlier.

Models of human numerosity processing typically contain an accumulator component.[7] My former student Marco Zorzi, now leader of one of the most imaginative number labs in Europe, and I proposed a neural network model of number processing in which we tested the idea that the system had 'inherited' an accumulator. In our studies and subsequent work by Zorzi with Carlo Umiltà and Ivilin Stoianov from the University of Padua, this approach was able to model human accuracy and timing data from numerosity comparison, symbolic calculation and number priming tasks very accurately.[8] Similar models but lacking the inherited component failed to model human data.

Overleaf is a sketch of our model as it was elaborated by Zorzi, Stoianov and Umiltà.

This model embodies a simple accumulator-type mechanism so that each item counted, for example in the dot patterns, adds a unit to the numerosity nodes. The five and three components are like the relative levels in two accumulators, and the eight component is the simple linear sum of the five and three components. In this model there is no explicit representation of noise, of 'scalar variability', that would be needed to make this model a closer

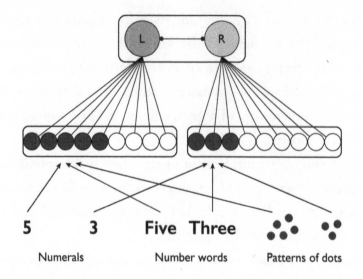

Figure 1. The basic model for number comparison. The 'numerosity code' straightforwardly represents number magnitude as the number of units activated within it – the equivalent of an accumulator. These nodes activate two possible responses (left or right) with a winner-take-all through competitive interactions (lateral inhibition). This model accounts for the basic response time data, comparing digits, words or dot patterns.[8]

approximation of accumulator addition. Instead, the decision process, rather than the representations, embodies the noise.

We also used it to model data we had previously collected on single-digit addition reaction times.

The model in Figure 2 turned out to be very accurate in modelling well-established reaction time human data for all single digit additions from 1 + 1 to 9 + 9, such as the 'problem-size effect' (problems with larger sums take longer answer), and the 'ties effect' (for the same sum, if the digits are the same, then the reaction time is shorter, e.g. 3 + 3 is quicker than 4 + 2).

Marco and I subsequently spent a very happy week at his

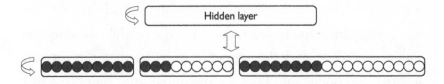

Figure 2. Learning addition in associative-memory network. In this model, the numerals 5, 3 and 8 are interpreted as numerosity nodes. The system cycles through the relationships among the 5, 3 and 8 until it settles on the familiar and correct state 5 + 3 = 8.[8]

house in the Dolomites working on a more elaborated version of the model, refining the fit to human data that we and others had collected earlier and demonstrating that models without the accumulator style numerosity code failed to fit these data. The paper, alas, is still languishing on our computers. Maybe it will see the light of day, eventually.

There are also neurons in the brains of macaque monkeys that do not act like accumulators. Rather, they respond maximally to particular numerosities. So, there will be a neuron that responds maximally to five objects, and it will respond less strongly to four or six objects. These were first identified by Andreas Nieder, now at the University of Tübingen, when he was working in Earl Miller's lab at MIT. He found neurons in both monkey frontal lobes and the parietal cortex in a comparable location to those in human parietal cortex.[9]

Computer models similar to AI and machine learning models propose an accumulator layer (called a 'summation' field) that feeds into a layer with elements that behave like Nieder's number neurons.[10] Another, but not dissimilar approach, calibrates levels in an accumulator and links the level to counting words.[11]

Zorzi, Stoianov and Umiltà used their model to explore

acquired dyscalculia, that is, an impairment to an adult's arithmetical competence due to brain damage. The model contained the numerosity code and the symbolic code, so that the network learned the connections between the numerosity codes (the five-dot string plus three-dot string to yield the eight-dot string). It also learned the connections among the symbols 5, 3 and 8. The trained network was 98 per cent correct on producing the correct sum when presented with the two addends, and showed the problem-size effect. The numerosity code or the symbolic code in the trained network was then randomly damaged by randomly eliminating 20 per cent, 50 per cent or 80 per cent of the connections between numerosity codes or symbolic codes. The result was very clear: damage to the numerosity code produced a far bigger impact on performance than damage to the symbolic code. Even elimating 20 per cent of the connections resulted in performance dropping by 60 per cent, whereas the effect of eliminating the symbolic connections had only a very small effect.[12] That is to say, how the network 'understands' arithmetic in terms of numerosities is critical; rote learning of the connections among uninterpreted symbols by itself is not a good idea, assuming, of course, that the model is a good representation of human addition fact memory.

This result relates indirectly to animal competence. The numerosity code in an accumulator type of system is critical at least to addition.

Developmental dyscalculia is different from the acquired type. It prevents the normal development of arithmetical competence, and is due to a 'core deficit' in something very similar to the numerosity code accumulators represented in the model. This means that sufferers find it very difficult to learn the addition facts, for example, because they don't have critical numerosity representations of the numbers.[13] It is something you are born with and persists into adulthood, and is quite independent of other cognitive abilities or disabilities, and is probably independent of the capacity for other aspects of mathematics. One can think of it as being rather like colour blindness: bad luck, a genetic anomaly, but it is possible to develop compensatory strategies, such as learning that the top traffic light is red, and the bottom light is green. At present, we know a lot about the genetics of colour blindness but not about the genetics of dyscalculia, except that there is a large genetic component to it in many cases. It is possible that some individual animals are also dyscalculic. For example, the average small fish – guppy or zebrafish – is quite good on numerosity discrimination and numerosity match-to-sample, but in our tests, some individual guppies seem consistently worse than others.[14] We are currently investigating the genetic bases of these differences.

Many creatures travel considerable distances to forage, to mate, to spawn or to overwinter. I have made the claim that many creatures have a cognitive map of their environment and that all creatures that move about in their environment need to carry out computations involving direction and distance, and that these computations involve maps, just as human navigators depend on compass, chronometers (to measure speed, time and hence distance) and a map to plot where they are in relation to where they have been and where they are going. Moreover, these computations are made through numbers. There is no other way.

The idea that rats have a cognitive map was first proposed in a wonderful paper by Edward Tolman in 1948. It was published in the leading psychology journal, *Psychological Review*, but as well as its ground-breaking proposal it was written in a personal and engaging way that I am sure would not be acceptable to the current editors of that distinguished organ.[15] For example, he introduces the experiments that

> were carried out by graduate students (or underpaid research assistants) who, supposedly, got some of their ideas from me. And a few, though a very few, were even carried out by me myself . . . In the typical experiment a hungry rat is put at the entrance of the maze . . . and wanders about through the various true path segments and blind alleys until he finally comes to the food box and eats.[15]

As the rat learns the maze, searching for the food reward becomes more efficient, avoiding known blind alleys, even when the rat starts from a different place. Information about the maze is received by 'the central office', which

> is far more like a map control room than it is like an old-fashioned telephone exchange. The stimuli, which are allowed in, are not connected by just simple one-to-one switches to the outgoing responses [stimulus-response connections beloved by behaviourists at that time]. Rather, the incoming impulses are usually worked over and elaborated in the central control room into a tentative, cognitive-like map of the environment. And it is this tentative map, indicating routes and paths and environmental relationships, which finally determines what responses, if any, the animal will finally release.[15]

He concludes with implications for humans of the various manipulations of mazes and rewards used with rats, in typical Tolmanian manner: 'My argument will be brief, cavalier, and dogmatic.' I doubt that an author today could get away with that, however distinguished.

> We must, in short, subject our children and ourselves (as the kindly experimenter would his rats) to the optimal conditions of moderate motivation and of an absence of unnecessary frustrations, whenever we put them and ourselves before that great God-given maze which is our human world. I cannot predict whether or not we will be able, or be allowed, to do this; but I *can* say that, only insofar as we *are* able and *are* allowed, have we cause for hope.[15]

It is no longer disputed that mammals have cognitive maps. We have a pretty good idea of how they plot direction using a sky compass or the earth's magnetic field, sometimes both. We have a very good idea of the neural mechanisms that implement it from the work of Nobel laureates John O'Keefe at UCL and Edvard Moser and Mai-Britt Moser at the Norwegian University of Science and Technology.

Rats are born with two brain systems to encode their cognitive map. *Place cells* in the hippocampus encode unique locations in unique environments.[16] *Grid cells* in the neighbouring entorhinal cortex fire when the animal moves in relation to external landmarks and provide direction and distance information for computing path integration.[17] Both types of cell encode different maps, and also local features: for example, food odours – 'food here'. These maps are continually updated, and the rat's brain may contain at one time multiple maps, at different scales and for different locales. Theoreticians are now modelling the activity of place and grid cells in relation to spatial orientation. How are they doing it? By programming digital computers – that is, numbers all the way down to zeros and ones.

The philosophers Blaise Pascal (1623–1662) and Immanuel Kant (1724–1804) proposed that the principles of space, time, motion and number are possessed by humans *a priori*, which we would now call 'innate' or inherited. As neuroscientists Stanislas Dehaene and Elizabeth Brannon declare, 'If Immanuel Kant or Blaise Pascal were born today, they would probably be cognitive neuroscientists!'[18]

The hippocampus and entorhinal cortex are found in mammalian brains, so the question arises as to whether non-mammalian species also possess cognitive maps. In Chapter 6 I argued that migrating birds' extraordinary route finding depended on having a cognitive map, and I suggested in Chapter 7 that this was also true for migrating sea turtles. It has been proposed that birds and reptiles have brain structures that are functionally equivalent to the mammalian hippocampus and cortex. So maybe it is not such a stretch to propose that these structures could embody a cognitive map. The dependence on route finding – path integration – in insects, however, is much more controversial. Invertebrate brains are very different even from reptilian brains. Of course, birds, rats and humans all descend from a common invertebrate ancestor, so perhaps it is not such a stretch to suppose that a neural structure in invertebrates has evolved into the mammalian mapping systems.

What we have that other animals don't have

In her fascinating imagining of the life of a Neanderthal clan, novelist Jean M. Auel notes one crucial cognitive difference between them and a five-year-old anatomically modern human, Ayla, they have rescued: 'Numbers were a difficult abstraction for people of the Clan to comprehend. Most could not think beyond three: you, me and another. It was not a matter of intelligence.'[19] But Ayla immediately grasped that the number of tallies and the number of fingers on them were the same, and could use her fingers to calculate the number of years until she became a woman. Her teacher, the greatest magician in all the clans, was 'rocked to the core. It was unthinkable that a child, a girl child at that, could reason her way to that conclusion so easily.' I don't know if Auel had read Bertrand Russell's (1872–1970) *Introduction to Mathematical Philosophy* (1919), but in it he wrote, 'It must have required many ages to discover that a brace of pheasants and a couple of days were both instances of the number 2: the degree of abstraction involved is far from easy.' Perhaps, then, Neanderthals hadn't managed to get to that degree of abstraction yet.

Of course, Auel's novel, *The Clan of the Cave Bear*, isn't science, though it is deeply informed by the science of that time, 1980, and actually anticipated the science to come. Auel's Neanderthals were not Haeckel's *Homo stupidus*, but skilful toolmakers, with a sophisticated gestural language, elaborate myths and rituals, but limited speech which she attributes to their vocal apparatus. They interacted, and occasionally interbred, with the Others, anatomically modern humans, which was unthinkable when she wrote, but genetic research not available in 1980 shows that this was indeed the case. Later evidence from La Pasiega Cave (Chapter 3) and other sites suggests that Neanderthals developed tallying, and they may have had counting words to go with their tallies. So I doubt that lack of counting skills explains the disappearance of our close cousins.

When humans are not allowed to count out loud, they behave just like other animals. One example I mentioned in Chapter 2, on humans, is when they have to press a key a given number of times while repeating 'the' as fast as possible. The distribution of their errors is remarkably similar to the mice we saw in Chapter 5 (though the mice are not saying 'the' repeatedly). The errors – pressing the key too often or not often enough – show 'scalar variability'. That is, the number and size of errors increase in proportion to the target number: the larger the number, the more and the larger the errors. However, when human participants are allowed to count verbally, the pattern of errors changes. Now the variability was proportional to the *square root* of the mean of the given target ('binomial variability').[21]

This makes sense. When verbal counting is suppressed by repeating 'the', the human relies on the accumulator alone, which has scalar variability (Chapter 1). When verbal counting is allowed, errors will arise when a count is missed or an object double-counted, and the larger the target number, the more opportunity there is for these errors, hence binomial variability.

Counting words provide humans with other advantages in addition to reducing counting errors. They enable more precise distinctions than is possible when comparing two accumulator levels for the sizes of larger sets, for example, 'thirty-three' vs 'thirty-four'. Second, they provide an efficient way of remembering the result of the count over the long term. You can remember the word 'thirty-four' that denotes a precise number, not a noisy accumulator level. Third, with the appropriate rules for generating new counting words, it is possible to count as high as is needed.

It has been argued by Noam Chomsky and his colleagues that humans are uniquely equipped with a computational mechanism for recursion and this enables us to use language with its indefinitely long sentences,[22] such as 'She swallowed the bird to catch the spider who wiggled and jiggled and tickled inside her and she swallowed the spider to catch the fly but I don't know why she swallowed the fly'. With this mechanism we can also create indefinitely large numbers, such as 'Two hundred million four thousand five hundred and sixty three'. Notice a particularity of English number word syntax: the 'and' only comes after the hundreds. There are in fact many particularities of number syntax, even compared with other quantity expressions. We can say *too many, very many, how many*, but not *too six, very six* or *how six*. The *many* examples are generated by the usual rules of English syntax, but the *six* examples are not and are thus just not grammatical. On the other hand, we can say *exactly six, less than six, almost six*, but not *exactly many, less than many, almost many*, again revealing a contrast between number syntax and the syntax for non-numerals.[23] And why can we say *thirty-seven hundred* but not *thirty hundred*? Nevertheless, we can make permanent records of numbers using counting words if we have a system for writing down words.

Counting words not only provide humans with important advantages – a way of accurately enumerating large sets and remembering the result of the count – they also have important disadvantages. They are name-value systems with distinct names for decades (ten, twenty, thirty) and for powers (hundred, thousand, million). This means that if there isn't a word for it, you're in trouble. This is a disadvantage that confronted Archimedes. He wanted to calculate how many grains of sand would fill the universe, and for that he would need a very large number indeed. The Greek of his time only had a word for 10,000 (10^4), a 'myriad'. As the greatest mathematician of antiquity, Archimedes saw a way through the problem. First he proposed a myriad myriads (10^8) which he called a unit of the first order; then there was a unit of the second order, $10^8 \times 10^8$. Then he defined 'periods' in which second order units became the *powers* of second-order units: 10^8 to the power of 10^8, and so on. Thus he invented a numeral system of multiples and powers six hundred years before the Indus Valley civilization and one thousand years before it was in widespread use in Europe.

Counting words don't make calculating easy. Even cultures with established and widely used counting words don't use these words for calculating. The Romans and Greeks, and the Incas, among others, used counting boards (Chapter 3). The Chinese and Japanese used the abacus. Counting boards and the abacus are not name-value systems, but *place-value* systems, just like the modern Hindu-Arabic numerals, with units, ten and hundreds in separate columns. This is why Fibonacci's (*c.*1170–*c.*1240/50) great book *Liber Abaci* (1202), which introduced these numerals to Europe, was so important.

Notation, notation, notation

'By relieving the brain of all unnecessary work, a good notation sets it free to concentrate on more advanced problems, and in effect increases the mental power of the race.'[24] So wrote the mathematician and philosopher A. N. Whitehead (1861–1947) in 1911.

Our familiar numerals were a massive step forward in relieving the brain of unnecessary work. Imagine what it was like for the numerate Roman citizen trying to calculate 325 × 47 using Roman numerals. I take this example from Graham Flegg's wonderful history of numbers through the ages:

> Try to multiply CCCXXV by XLVII . . . The first problem that arises is that XLVII cannot be decomposed into parts X + L + V + I + I, because the notation XL is subtractive. One can try writing XXXX instead of XL and try computing the product of CCCXXV and XXXXVII by multiplying every single component C or X or V contained in the first factor by component X or V or I of the second factor. This method will involve 42 (6 × 7) single multiplications, followed by the addition of the results.[25]

No wonder traders all over Europe sent their sons to places where they could learn the new numbers. (There is no evidence that daughters were sent.) In Italy, they were called *scuole d'abbaco* run by a *maestro d'abbaco*. Leonardo da Vinci attended one in Florence. One city famous for these *scuole* was Venice. The old post office by the Rialto Bridge, now a fancy department store, was originally the *Fondaco dei Tedeschi*, where the sons of German traders learnt the new numbers and double-entry bookkeeping.[26]

In her subtle and profound story 'The Masters', Ursula Le Guin (1929–2018) imagines a world in which the only people initiated into the rites and mysteries of the Lodge are able to do arithmetic, and every business will need a Master from the Lodge to help solve everyday business problems. In Le Guin's grey, rain-sodden world, a heretical Master shows the hero how to calculate using a symbol for *nothing* to create a positional numbering system with a base of ten. 'Numbers are the heart of knowledge, the language of it,' says the Master.[27]

Positional notation is not just useful for written calculations. You and I and all the great calculators I described in Chapter 2 use the notation for mental calculation. They certainly don't use Roman, Greek or Hebrew name-value notations. Abacus masters and abacus competitors translate numerals into a mental abacus, also positional, and use that to carry out the calculation.

As we saw in Chapter 3, humans in early historical times and prehistory recorded the results of their counting independently of the counting words they may have used in speaking, by tallying on bones, stones and cave walls.

Education, education, education

The other thing that we have but other creatures don't is education, which enables inventions, such as positional notation, to be diffused from their inventors to learners. Education of this sort rarely happens in the animal world and seems to be confined to tool use and foraging.

One can think of animal training experiments as a kind of education with a teacher. Typically training animals in the lab – or for entertainment – is a slow and laborious process precisely because the animal subjects are not doing what comes naturally. For example, the celebrated chimpanzee Ai (see p. 121) took 1821 trials over four hours to learn the set sizes associated with the first number symbols ('1' and '2').[28] Cantlon and Brannon's study of 'basic math in monkeys and college students' demonstrates the difference between monkeys and humans. It took the monkeys 500 trials to get better than chance on problems such as 1 + 1 = 2, 4 or 8.[29] Mechner's original study with rats reports that 'after protracted training' they could press a key a target number of times.[30] Meck and Church's classic accumulator study with rats learning to recognize the number of tones reports fifteen days of training before two days of testing. The remarkable parrot Alex underwent thirty *years* of training for speech and numbers, with some kind of practice almost every day.

The problem of zero

I have rather tried to avoid this issue, since it comes with one big problem. In his classic history, *Number: The Language of Science* (1930), Tobias Dantzig (1884–1956) wrote that 'The discovery of zero will always stand out as one of the greatest single achievements of the human race'.[31] It was a hard-fought achievement that first saw the light around 600 CE when developed by brilliant mathematicians, or perhaps one particularly brilliant mathematician, in the Indus Valley. And it took another 600 years for zero to be adopted in Europe following the publication of Fibonacci's *Liber Abaci* (see p. 292). We now know that actually the Maya invented a symbol for zero four hundred years before the Indus Valley mathematicians. The great Greek mathematicians, including Euclid and Archimedes, had no symbol for zero. Surely, then, if even the great Greek mathematicians were without a zero, and it took a great discovery to find it, no non-human animal could possess this concept.

And yet, Andreas Nieder and his colleagues at the University of Tübingen published a paper in 2021, showing that crows can distinguish zero from other numbers of objects in a way that suggests they treat zero objects as just another number.[32] Not only that, but they found neurons in a region of the crow's brain that responds to zero objects. These neurons were in the same part of the brain that responded to other numbers and seemed to be tuned in the same way so that they fired maximally to the preferred number but also to a lesser extent to adjacent numbers. The 'three neuron' would also fire to two and four objects in a mathematically predictable way. So the zero neuron would fire most to a display with no objects (dots on a screen) – the empty set – but a bit also to a display with one object, and a tiny amount to two, three and four objects.

This study was not the first to show that other creatures respond distinctively to zero objects. Tetsuro Matsuzawa at Kyoto University's Primate Research Institute showed that a chimp that has learned to accurately match the symbols 1 to 9 with the number of randomly arrayed dots can learn to match the symbol '0' to a display of no dots.[33]

Elizabeth Brannon's lab at Duke University tested for zero in both preschoolers[34] and rhesus monkeys.[35] After they had learned to select the smaller of two set sizes in order to receive a reward, they were more likely to select the empty set over larger set sizes in a test, indicating that they appreciate that zero is the smaller value on their internal scale of numerical values.

It turns out that ingenious experimenters can show that creatures with very tiny brains, such as honeybees with fewer than 1 million neurons (we have 86+ billion neurons), seem to have a sense of zero. An Australian team led by Adrian Dyer has elegantly demonstrated that even insects order zero along the numerical continuum.[36] They first trained the bees to pick one of the two displays with the most objects, rewarding them with a tasty sucrose reward and with a bitter quinine solution (punishment) if they selected the smaller numerosity. They also trained other bees in the same way, but this time to choose the display with fewer objects. Then they introduced novel displays, including displays with no objects. The bees trained to choose more did not choose the empty set; those trained to choose fewer would choose the empty set. In both cases, the bigger the difference, the more likely was the bee to make the correct choice. This is Weber's Law all over again. The authors conclude: 'Bees could subsequently extrapolate the concept of less than to order zero numerosity at the lower end of the numerical continuum.'

Now, perhaps it's not surprising that animals and humans can distinguish something from nothing, and perhaps it is easier for them and us to distinguish lots of things from nothing than one thing from no things. That's not why it took humans more than a hundred thousand years to invent zero.

Stone Age humans, including Neanderthals, marked bones, stones, cave walls and presumably sticks, though these no longer survive, using a tally system: one incision on a bone, or dab of ochre on the wall, for each object counted. For the empty set, there would be no mark. Suppose our Palaeolithic hunter was representing the number of deer killed today, and had killed only the empty set of deer, no mark would be added to the tally.

Zero to represent the empty set is only one way to think about it. The reason why it took humans so long to invent a symbol for it is because the symbol is part of a much wider and more radical way of representing numbers – the place-value system. The zero symbol only makes sense as part of the system. Human verbal counting from earliest times has not been place-value but name-value. For each power of ten we have separate words or phrases: *ten, hundred, thousand, ten thousand, million* and not *one zero, one zero zero, one zero zero zero* . . . Children have to learn a procedure to translate the words into the digit symbols, and that takes a while. They go through a phase of getting it wrong because it is actually quite complicated. So for example they learn that 'a hundred' is 100 in place-value notation, and for a while may write 'a hundred and three' as *1003*, or 'a hundred and twenty-three' as *10023*. The rule for overwriting the rightmost zeros takes time to master.

The brain seems to have a special place for this procedure. Lisa Cipolotti, a neuropsychologist at the National Hospital for Neurology in London, examined one neurological patient, D.M., who for a while seemed to regress when writing to dictation four-, five- and six-digit numbers.[37] All his errors involved the insertion of additional zeros. For example, when asked to write 'three thousand two hundred', he wrote 3000,200; for 'twenty-four thousand one hundred and five', he wrote 24000,105. However, he invariably read the numerals correctly and after two days, he began to realize his error and managed to start writing them correctly.

With Italian colleagues from the University of Padua, we found a particular region in the right hemisphere (the insula) that had been damaged in twenty-two patients, and was linked to problems reading and writing numbers with zero in them. For example, 70,002 was read as 'seven thousand and two'; 'ten thousand and fifty' was written as 100,050; 'one hundred thousand and three' was written as 10003.[38]

These patients help to illustrate the very important property of zero, namely its essential role in the development of place-value notation. To learn how to use zero in the notation is difficult for children, and has been hard for humans. *Liber Abaci* showed merchants how to make calculations, as well as keeping and agreeing accounts, simpler. This quite literally changed the world, but it was a long time coming.[39]

What we don't know

In their review of the numerical abilities of non-human species, Davis and Pérusse head their summary 'from backwater to mainstream of comparative psychology' to indicate that the study of the numerical abilities of non-human species has now become a very large field of enquiry, with many more species investigated in many more natural and laboratory settings. And they conclude that there is clear evidence that animals, 'although not naturally attuned to numerical stimuli, can become responsive to such events under supportive environmental condition'.[40]

Some animal groups have received more investigation than others. Most of the studies have been on our nearest relatives. Fish are catching up, but still have a long way to go. Even among fish, only a few species have been tested. Guppies (*Poecilia reticulata*) are by far the favourite freshwater fish, in aquariums and in the science of numerical cognition. In my own experience, guppies are easier to train to make numerical choices than zebrafish (*Danio rerio*). Zebrafish are widely used by scientists interested in the genomic basis of cognitive abilities because their genome has been sequenced and transgenic lines are relatively easily available to test the effects of individual genes. The larval form is transparent, which makes brain imaging possible in some transgenic strains. The biggest jump has been in studies of invertebrates.

One other trend since the famous Royal Society meeting in 2017 has been the thirty-seven substantial human studies that reference the findings in other animals, including in studies of economics, media impact and education.

My claim is that all animals need to be able to read the language of the universe, but only a few of the perhaps 10 million species have been tested. And that's just animals. It is claimed that at least some plants can count. Charles Darwin called the Venus flytrap *Dionaea muscipula* 'the most wonderful plant in the world'. Darwin observed that the hungry plant develops a red-coloured inner trap to lure flies, and that the trap shuts only when there were two consecutive touch events. German plant scientists Rainer Hedrich and Erwin Neher have investigated the method by which plants do this. There are touch-sensitive mechanoreceptors, rather like hairs, on the surface of the trap, which generate an electrical signal. When two of these occur within about thirty seconds, the trap closes and the insect is consumed by the plant's digestive juices. This means that the Venus flytrap has the ability to remember the touch for at least thirty seconds, and count that a second touch has occurred.[41]

As well as plants, are there other life forms that may be able to count among one *trillion* species of microorganisms on Earth (more than the number of stars in the Milky Way galaxy)? Can bacteria count, for instance? This is an area in which I am particularly ill informed, and I was pointed to it by an intriguing description of something like counting in bacteria by Andreas Nieder in his book *A Brain for Numbers*.[42] Bacteria are highly gregarious organisms and some behaviours that are important for reproductive success they will only do in company, because company is important for reproduction. This is called 'quorum sensing'. The marine bacterium *Vibrio fischeri* was found to have a gene, *lux*, that enables bioluminescence, rather like fireflies.[43] It turns out that the bacterium only lights up when a certain number of other bacteria of its own kind are nearby, and then they all light up together. *Vibrio fischeri* bacteria produce a molecule that enables others to recognize their presence. And, it turns out, quorum sensing may be a route to therapies by suppressing this mechanism – 'quorum quenching' – to stop harmful bacteria reproducing.[43] However, it seems to take the presence of at least a million others for bioluminescence to occur and I rather doubt that these single-celled organisms can count that high. In any case, there is yet no evidence that either the Venus flytrap or the bacteria satisfy my second criterion, namely that they can compute operations that are isomorphic with arithmetical operations. I wait to be proved wrong on this.

Another missing piece of the puzzle is ecology. We still know relatively little of how counting and calculation are used in the wild. We know a lot about the numerical capabilities of rats and mice in the lab, but little about how they use these capabilities going about their everyday lives.

By contrast, we know a lot about the computations needed for real-life navigation in ants, bees, reptiles and birds, but these actual computational abilities have only been intensively studied in ants, bees and the mantis shrimp. We know how they evaluate their own direction using a sky compass and/or the earth's magnetic field, and how they estimate distance travelled in each direction. Desert ants count their steps, bees count landmarks, and probably all these navigators use a sense of their own movement and its duration to compute the distance travelled in each direction – path integration – so that they know where they are and, if necessary, how to get home by the shortest route.

Although these numerical computations are necessary for navigation, there are still many unanswered questions about how our brains or the brains of other creatures represent numerosities.

Roitman and colleagues identified neurons in the parietal lobes of monkeys that encode 'the total number of elements within their classical receptive field in a graded fashion, across a wide range of numerical values (2–32)'.[44] Moreover, modulation of neuronal activity by visual quantity developed rapidly, within 100 msecs of stimulus onset, and was independent of attention, reward expectations or stimulus attributes such as size, density or colour. The responses of these neurons resemble the outputs of 'accumulator neurons' postulated in computational models of number processing. Broadly speaking, the neuronal responses are proportional to the amount of input. That's not really mysterious, though the details need to be spelled out. Numerical accumulator neurons may provide inputs to neurons encoding specific cardinal values, such as 'four', that have been described in previous work such as Nieder's 'number neurons', which are broadly tuned to particular numerosities.[42] It is probably not a coincidence that accumulator neurons and Nieder neurons are close neighbours in the monkey brain.

There is still the question of how individual neurons of the Nieder type actually encode a number. At the moment, we don't know.

Now both the Nieder neurons and the accumulator neurons assume that there is something special about the brain's response to numerosities. For example, we could be born with these neurons, which are just waiting for the animal to begin experiencing the world to start working.

There is another approach very much in line with modern AI and which is called 'deep learning'. These are computational networks that learn using algorithms originally developed by Geoffrey Hinton, now at the University of Toronto and at Google, and his colleagues. Marco Zorzi's lab at the University of Padua has modelled the numerosity comparison task by letting the network learn without any special information about numerosities. This approach is called 'unsupervised learning', which is a kind of learning that uses the statistical regularities in the data, which in this case is sensory input. This is the kind of model used for face recognition software to assess whether these two, three or more faces are the same. The question is, will the network learn whether these two, three or more displays are the same, and can this be used to select the larger of two sets of dots? Zorzi and colleagues made this task particularly difficult in two ways. First, they threw in all the visual features that make the task difficult by varying the size, spacing and arrangement of the dots, so that, for example, the array of dots covering the larger area or the highest density may have fewer dots. It turned out that numerosity is spontaneously encoded just by looking at the data, and even in the absence of a task to discriminate the numerosities of sets. Second, they asked forty volunteers to do the same task as the computer. The results showed that the computer network could learn to model human performance accurately, and moreover, like the humans, the most important dimension of the task was actually the numerosity. That is, the computer automatically extracted numerosity from the displays without being specifically trained to do so.[45] This is very different from what Zorzi called, in a recent email to

me, the 'prehistoric modelling work' that is depicted in Figures 1
(p. 282) and 2 (p. 283).

Despite the brilliance and originality of this work, it is not
clear to me where the actual numbers are in the network.

The metaphysical rabbit hole

In a forthcoming paper about navigating real and scientific waters,
Gallistel summarizes his position on cognitive maps thus:

> The cognitive map hypothesis entails that there be strings
> of numbers in neural tissue on which neural machinery
> operates. However, some neuroscientists – actually, many,
> indeed most – may well object. What could a string of
> numbers look like in neural tissue? And what could the
> machinery that operates on this neurobiological fantasy
> possibly look like? Is it not just as implausible to imagine
> that there are numbers in neural tissue as to imagine that
> there are arrows? Not at all! However, to realize this we
> must be clear that we are talking about numbers as a com-
> puter scientist understands them not about the Platonic
> numbers of mathematicians, logicians and philosophers.
> We must not go down the metaphysical rabbit hole of
> worrying what a number really is. (In arguing this thesis,
> I find that I cannot keep most neuroscientists and many
> cognitive scientists out of that hole.)

Despite what Gallistel says, we, as scientists, have to go down
into the rabbit hole occasionally. The question is how to clamber
out of it. The problem is, and here I follow the philosopher

Marcus Giaquinto, 'Numbers cannot be seen, heard, touched, tasted, or smelled; they do not emit or reflect signals; they leave no traces.'[46] They are, as I said in Chapter 1, abstract: the number three can truly refer to any set of three things. That is, threeness is a property of a set, and an abstract property at that. This leads to an epistemological problem. Abstract objects are not objects in the world, and therefore cannot have causal relations with anything, including us. This has been realized from at least the time of Plato (*c*.428–*c*.348 BCE). He argued that we have a special kind of intuition to perceive these objects, which exist not in the real world or in the world of our thoughts, but in a 'third world' of abstract objects including mathematical objects. This intuition is a kind of memory of a previous life, as Socrates (*c*.470–399 BCE) seeks to demonstrate with an illiterate slave boy in his dialogue *Meno*, who, after skilful Socratic questioning, remembers facts of geometry, something he had never learned in this life.

Now you might think that this reminiscence view of numbers and other abstract objects is absurd, but many mathematicians are Platonists in this sense. Although they don't talk about remembering a new property of prime numbers, they talk about *discovering* mathematical truths in the world of prime numbers, and other abstract objects, as if they were discovering a new continent or a new property of the element oxygen. Kurt Gödel (1906–1978), one the greatest logicians of the twentieth century, is a prime (apologies) example of this kind of mathematician.

So, for example, one could say that Euclid's proof that there were infinitely many primes (*Elements*, Book IX, Proposition 20) was a discovery of the infinitude of primes. He certainly didn't discover it by counting objects in the world.

However, if threeness is a property of all sets of three things, how can we know this if threeness has no causal role in the world? Giaquinto's argument is that we need to rethink the causal theory of knowledge. We actually know lots of abstract things. In fact, most of what we know are abstract things through experience of their exemplars or instances, which do indeed have causal properties. All readers of this book know the letters of the English alphabet, but they know 'A' in an abstract way since they can recognize it irrespective of its size, colour, font and case, or whether it is printed or handwritten. Theories of human reading therefore include a type of representation of 'A' called a 'grapheme', rather than a letter, indicating that this is a category denoting all versions of the letter. The same is true of spoken words. We know the word 'cat' irrespective of whether it is spoken by a male or female, high or low pitch, accent, volume and so on. Similarly, we know the tune *Happy Birthday* whether it is sung, played on a banjo or performed by a full orchestra. How we actually come to know these objects is a complicated matter for cognitive science rather than philosophy.

Numerosity, according to Giaquinto, is a property of the physical and perceptible world. I gave some examples in Chapter 1. These are real properties of the real world. Things would be very different if these numbers changed – for example, if we had three arms and three eyes.

Another version of this idea has been proposed by the MIT physicist Max Tegmark, in his book *Our Mathematical Universe*.[47] He argues that the physical universe is not just *described by* mathematics, but *is* mathematics. Mathematical existence, for him, equals physical existence, and all structures that exist mathematically exist physically as well. So we are back to Pythagoras, who allegedly said, 'All is numbers.' There is an obvious problem with this position, as others have pointed out. The structure of the natural numbers entails infinity but there are not infinitely many things in the universe. The focus on mathematical *structures* may be a way out.

Anyway, Giaquinto tries to solve the epistemological problem by arguing that we know about abstract numerosities from experience of their instances.

But hold on a minute. That can't be the whole story. The brain must be prepared to identify the numerosity of the sets it experiences: it must be able to recognize that these two sets have the same numerosity or that they have different numerosities. For this, the brains must be equipped with an accumulator system or something equivalent to it, and the whole system, including the selector, must be able to generalize across different instances, and that is the basis of abstraction.

As I argued in Chapter 1, abstraction is not all or none, but operates in degrees. Humans can generalize across sets of any types of objects, but other species may be able to generalize across sets of just one or just a few types of adaptively relevant objects. The system operates automatically and mandatorily: if there is a set of three apples, its threeness is automatically registered in the brain, and this registration cannot be turned off. The animal may or may not do anything with this information. That is for other brain systems to decide.

The brain does not come into the world as a 'tabula rasa' – a blank slate. The starter kit for learning arithmetic must have some built-in mechanisms that it has inherited from distant or less distant ancestors.

The idea of a built-in mechanism for registering numerosities, and indeed one that the animal inherits, is really part of a widespread standard practice in nature. We know that creatures from birth represent objects and events in the environment numerically. We have seen that human infants respond to numerosities in the first days of life (Chapter 2); newly hatched chicks (Chapter 6) and guppies (Chapter 8) also respond to numerosities in the environment without any training or indeed much experience.

We also know that these built-in mechanisms are critical for recognizing and representing all kinds of objects and events in the environment. To tell whether two objects have the same or different colour it is necessary to have a working inherited built-in colour vision system. Individuals with colour blindness – colour vision deficiency – may not be able to do this. In this case, the mechanisms in the eye and brain (actually the colour-sensitive cones in the retina are really part of the brain) are well understood. Even the genes involved in building the colour vision system are well understood.

Animals may have different colour vision from us. Bulls, famously, do not see red: they charge the matador's cape because it is moving. Dogs and cats have much poorer colour vision than us. Dogs possess two classes of cone pigments, one sensitive to long/medium-wavelength light (red/green) and the other sensitive to short-wavelength light (blue), whereas our retinas contain three classes of cone pigments (red, green and blue). That means dogs can't distinguish between red and green, like a common form of human colour vision impairment.[48] Matsuzawa showed that the chimpanzee Ai had colour vision very similar to humans,[28] and that seems to be true of other primates. Bees too have excellent colour vision, and in fact they can see colours we cannot. They use ultraviolet light in the sky – invisible to us – to locate the solar azimuth in partly overcast conditions (Chapter 9). The mantis shrimp sees the polarization of sunlight, again invisible to us, to help plot its route to and from a food source (Chapter 9).

Similarly, we have a built-in system for representing pitch. Again, individuals lacking this system suffer from *congenital amusia*, a lifelong disorder of music processing due to impaired pitch perception and memory. It also affects hearing pitch in speech and recognizing different intonations, such as the rising intonation in questions.[49] For these sufferers, recognizing different examples of a tune is near impossible.

As I have mentioned, a small proportion of humans, around 5 per cent, have difficulty in recognizing the numerosity of even quite small sets, and this results in serious problems in learning everyday arithmetic usually called 'dyscalculia'.[50] Although we have yet to discover the genetic basis for this disability, we do know from twin studies that it is inherited in many, but not all, cases. Dyscalculia implied that the other 95 per cent of us possess an efficient system for recognizing the numerosity of sets, and that this forms the basis of the development of our arithmetical abilities. Without it, they have trouble learning arithmetic in the normal way. We don't yet know whether, like colour blindness, it does not yield to intervention, but only to compensatory strategies; or whether it is treatable with the right interventions applied early enough. Much of my current research is focused on these two issues: the genetic basis of dyscalculia, and the best type of intervention for dyscalculic kids starting school.

Evolution of counting

While many scientists have been skeptical of claims for the numerical abilities of non-human species, followers and colleagues of

Darwin, such as George Romanes (1848–1894) and William Lauder Lindsay (1829–1880), were convinced that 'lower animals' had inherited a 'faculty of numeration'.[51]

Given the definition of counting I have been using throughout the book, Figure 3 summarizes the evidence for each group of animals. The groups correspond roughly to taxonomic classes, apart from amphibians and reptiles which are distinct classes. As all classes descend from a common ancestor, it is possible, and I would say necessary, that this common ancestor, whatever it was, could also count because all animals must to a greater or lesser extent understand the language of the universe if they are to thrive. Of course, there could be exceptions in this tree – animals

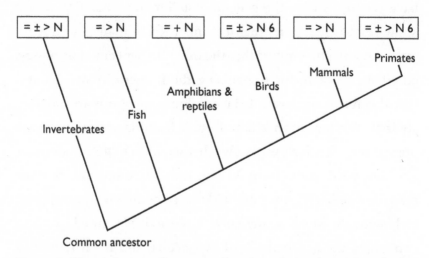

Figure 3. The evolution of counting based on the evidence described in the book. Tasks successfully achieved in properly controlled studies:

=   numerosity match-to-sample,
±   can add and/or subtract,
>   can select the larger (or smaller),
N   can carry out navigation computations,
6   can use numerals.

that cannot count – but those that have been tested show at least some counting ability in the lab, often after arduous training.

Where it is known, I have described how counting abilities are used spontaneously in the wild, and how these abilities help the individual animal forage, avoid death, and reproduce.

Table 1 is a brief and very rough summary of the evidence for the role of counting in the lives of these creatures. One reason why we know that counting is adaptive for food is that most lab experiments and lab-like experiments in the field have used food as a stimulus or as a reward. I haven't included navigation for two reasons. First, almost all animals navigate for food or sex or to avoid death. Second, although much is known about how

Group	Sex	Food	Death
Invertebrates	X	X	?
Fish	X	X	X
Amphibians and Reptiles	X	X	?
Birds	X	X	X
Mammals	?	X	X
Primates	?	X	X

Table 1. Evidence for the tested adaptive value of counting for the different classes of animals. 'Sex' means that counting enhances reproductive success; 'Food' means counting enhances foraging success; 'Death' means that counting enhances survival; and 'Navigation' means that numerical computation is deployed in travel to and from foraging or home sites. 'X' denotes evidence described in the book that shows a species counts for these purposes. '?' denotes that the role of counting for this purpose has not been properly established. Of course, the categories are overlapping: animals migrate for food and sex, and foraging supports migration. Survival enables sex and reproduction. Not all species in a class have been shown to count for the purposes in the columns.

animals use map and compass to navigate, I am not convinced we really know how they actually carry out the numerical computations needed for path integration.

The evolution of counting from the Cambrian explosion 600 million years ago to our present numerical sophistication, from first arthropods to positional notation, from fish choosing the larger shoal to investors choosing the best interest rate – these are part of a continuous historical process. It is clear from the studies of other animals and from the earliest members of genus *Homo* that we don't need counting words to count; we don't even need positional notation to calculate; but, as old Mr Locke said, these 'conduce to our well reckoning'.

REFERENCES

Chapter 1

1 Richardson, N. We're not talking to you, we're talking to Saturn. *London Review of Books* **42**, 23–26 (2020).

2 Giaquinto, M. Philosophy of number, in *The Oxford Handbook of Numerical Cognition* (eds R. Cohen Kadosh & A. Dowker), 17–31 (Oxford University Press, 2015).

3 Gallistel, C.R. Animal cognition: The representation of space, time and number. *Annual Review of Psychology* **40**, 155–189 (1989).

4 Gelman, R. & Gallistel, C.R. *The Child's Understanding of Number* (Harvard University Press, 1986; originally published 1978).

5 Locke, J. *An Essay Concerning Human Understanding*, ed. J.W. Yolton (J.M. Dent, 1961; originally published 1690). Book II, Chapter XVI.

6 Kahneman, D., Treisman, A. & Gibbs, B.J. The reviewing of object-files: Object specific integration of information. *Cognitive Psychology* **24**, 174–219 (1992).

7 Meck, W.H. & Church, R.M. A mode control model of counting and timing processes. *Journal of Experimental Psychology: Animal Behavior Processes* **9**, 320–334 (1983).
Meck, W.H., Church, R.M. & Gibbon, J. Temporal integration in duration and number discrimination. *Journal of Experimental Psychology: Animal Behavior Processes* **11**, 591–597 (1985).

8   Dehaene, S. & Changeux, J.-P. Development of elementary numerical abilities: A neuronal model. *Journal of Cognitive Neuroscience* **5**, 390–407 (1993).

9   Whalen, J., Gallistel, C.R. & Gelman, R. Nonverbal counting in humans: The psychophysics of number representation. *Psychological Science* **10**, 130–137 (1999).

10   Feigenson, L., Dehaene, S. & Spelke, E. Core systems of number. *Trends in Cognitive Sciences* **8**, 307–314 (2004). Carey, S. Where our number concepts come from. *Journal of Philosophy* **106**, 220–254 (2009).

11   Mandler, G. & Shebo, B.J. Subitizing: An analysis of its component processes. *Journal of Experimental Psychology: General* **11**, 1–22 (1982).

12   Balakrishnan, J.D., and Ashby, F.G. Subitizing: Magical numbers or mere superstition? *Psychological Review* **54**, 80–90 (1992).

13   Cantlon, J.F. & Brannon, E.M. Shared system for ordering small and large numbers in monkeys and humans. *Psychological Science* **17**, 401–406 (2006).

14   Piazza, M., Mechelli, A., Butterworth, B. & Price, C.J. Are subitizing and counting implemented as separate or functionally overlapping processes? *NeuroImage* **15**, 435–446 (2002).

15   Cai, Y. et al. Topographic numerosity maps cover subitizing and estimation ranges. *Nature Communications* **12**, 3374 (2021).

16   Brannon, E.M. & Merritt, D.J. in *Space, Time and Number in the Brain* (eds Stanislas Dehaene & Elizabeth M. Brannon), 207–224 (Academic Press, 2011).

17   Brannon, E.M., Wusthoff, C.J., Gallistel, C.R. & Gibbon, J. Numerical subtraction in the pigeon: Evidence for a linear subjective number scale. *Psychological Science* **12**, 238–243 (2001).

18   Karolis, V., Iuculano, T. & Butterworth, B. Mapping numerical magnitudes along the right lines: Differentiating between scale and

bias. *Journal of Experimental Psychology: General* **140**, 693–706 (2011).

19 Izard, V. & Dehaene, S. Calibrating the mental number line. *Cognition* **106**, 1221–1247 (2008). Siegler, R.S. & Opfer, J.E. The development of numerical estimation: Evidence for multiple representations of numerical quantity. *Psychological Science* **14**, 237–243 (2003).

20 Hollingworth, H.L. The central tendency of judgment. *Journal of Philosophy, Psychology, and Scientific Methods* **7**, 461–469 (1910).

21 Davis, H. in *The Development of Numerical Competence: Animal and Human Models* (eds S.T. Boysen & E.J. Capaldi) (LEA, 1993).

22 Koehler, O. The ability of birds to count. *Bulletin of Animal Behaviour* **9**, 41–45 (1950).

23 Bahrami, B. et al. Unconscious numerical priming despite interocular suppression. *Psychological Science* **21**, 224–23 (2010). Dehaene, S. et al. Imaging unconscious semantic priming. *Nature* **395**, 597–600 (1998). Koechlin, E., Naccache, L., Block, E. & Dehaene, S. Primed numbers: Exploring the modularity of numerical representations with masked and unmasked semantic priming. *Journal of Experimental Psychology: Human Perception and Performance* **25**, 1882–1905 (1999). Naccache, L. & Dehaene, S. Cerebral correlates of unconscious semantic priming. *Médecine Sciences* **15**, 515–518 (1999).

24 Parsons, S. & Bynner, J. *Does numeracy matter more?* (National Research and Development Centre for Adult Literacy and Numeracy, Institute of Education, 2005).

25 Gross, J., Hudson, C. & Price, D. *The long term costs of numeracy difficulties* (Every Child a Chance Trust, KPMG, London, 2009).

26 Bynner, J. & Parsons, S. *Does numeracy matter?* (The Basic Skills Agency, 1997).

27 Bevan, A. & Butterworth, B. *The responses to maths disabilities in the classroom* (2007). http://www.mathematicalbrain.com/pdf/2002 BEVANBB.PDF

28 Franklin, J. Counting on the recovery: The role for numeracy skills in 'levelling up' the UK (2021). https://www.probonoeconomics. com/news/press-release-covid-job-losses-disproportionately-impact-people-with-low-numeracy-skills

29 Smith, S.G. et al. The associations between objective numeracy and colorectal cancer screening knowledge, attitudes and defensive processing in a deprived community sample. *Journal of Health Psychology* **21**, 1665–1675 (2014).

30 Butterworth, B. *Dyscalculia: From science to education* (Routledge, 2019).

31 Beddington, J. et al, editors. *Foresight Mental Capital and Wellbeing Project: Final Project Report* (Government Office for Science, 2008).

32 Cockcroft, W.H. *Mathematics Counts: Report of the Committee of Inquiry into the Teaching of Mathematics in Schools under the Chairmanship of Dr W H Cockcroft* (HMSO, 1982).

33 *Mathematics learning in early childhood: paths toward excellence and equity* (National Research Council Center for Education, Division of Behavioral and Social Sciences and Education, 2009).

34 OECD. *The high cost of low educational performance: The long-run economic impact of improving educational outcomes* (2010).

Chapter 2

1 Butterworth, B. in *The Cambridge Handbook of Expertise and Expert Performance* (eds K.A. Ericsson, R.R. Hoffmann, A. Kozbelt & A.M. Williams), 616–633 (Cambridge University Press, 2018).

2   Smith, S.B. *The Great Mental Calculators: The Psychology, Methods, and Lives of Calculating Prodigies* (Columbia University Press, 1983).

3   Hunter, I.M.L. An exceptional talent for calculative thinking. *British Journal of Psychology* **53** (1962).

4   Hardy, G.H. *A Mathematician's Apology* (1940; Cambridge University Press, 1969).

5   https://www.youtube.com/watch?v=JawFocv5oLk

6   https://www.youtube.com/watch?v=_vGMsVirYKs

7   Binet, A. *Psychologie des grands calculateurs et joueurs d'échecs* (Hachette, 1894).

8   Horwitz, W.A., Deming, W.E. & Winter, R.F. A further account of the idiot savants: Experts with the calendar. *American Journal of Psychiatry* **126**, 160–163 (1969).

9   Scripture, E.W. Arithmetical prodigies. *American Journal of Psychology* **4**, 1–59 (1891).

10 Bahrami, B. et al. Unconscious numerical priming despite interocular suppression. *Psychological Science* **21**, 224–233 (2010).

11 Konkoly, K.R. et al. Real-time dialogue between experimenters and dreamers during REM sleep. *Current Biology*, doi:10.1016/j.cub.2021.01.026.

12 Fuson, K.C. *Children's Counting and Concepts of Number* (Springer, 1988).

13 Gelman, R. & Gallistel, C.R. *The Child's Understanding of Number* (Harvard University Press, 1978; 1986 edn).

14 Fuson, K.C. & Kwon, Y. in *Pathways to Number: Children's Developing Numerical Abilities* (eds J. Bideaud, C. Meljac & J.P. Fisher) (LEA, 1992).

15 Miura, I.T., Kim, C.C., Chang, C.-M. & Okamoto, Y. Effects of Language characteristics on children's cognitive representation of

number: Cross-National comparisons. *Child Development* **59**, 1445–1450 (1988).

16 Miura, I.T., Okamoto, Y., Kim, C.C., Steere, M. & Fayol, M. First graders' cognitive representation of number and understanding of place value: Cross-national comparisons: France, Japan, Korea, Sweden, and the United States. *Journal of Educational Psychology* **85**, 24–30 (1993).

17 Piaget, J. *The Child's Conception of Number* (Routledge & Kegan Paul, 1952).

18 Carey, S. Where our number concepts come from. *Journal of Philosophy* **106**, 220–254 (2009).

19 Núñez, R.E. Is there really an evolved capacity for number? *Trends in Cognitive Sciences* **21**, 409–424 (2017).

20 Bowern, C. & Zentz, J. Diversity in the numeral systems of Australian languages. *Anthropological Linguistics* **54**, 133–160 (2012).

21 Seidenberg, A. Ritual origin of counting. *Archive for the History of Exact Sciences* **2**, 1–40 (1962).

22 Locke, J. *An Essay Concerning Human Understanding*, ed. J.W. Yolton (J.M. Dent, 1961; originally published 1690). Book II, Chapter XVI.

23 Aikhenvald, A.Y. *Languages of the Amazon* (Oxford University Press, 2011).

24 Pica, P., Lemer, C., Izard, V. & Dehaene, S. Exact and approximate calculation in an Amazonian indigene group with a reduced number lexicon. *Science* **306**, 499–503 (2004).

25 Hale, K. in *Linguistics and Anthropology: In Honor of C.F. Voegelin* (eds M.D. Kinklade, K. Hale & O. Werner) (Peter de Ridder Press, 1975).

26 Wassmann, J. & Dasen, P.R. Yupno number system and counting. *Journal of Cross-Cultural Psychology* **25**, 78–94 (1994).

27 Kendon, A. *Sign Languages of Aboriginal Australia: Cultural, Semiotic and Communicative Perspectives* (Cambridge University Press, 1988).

28 Epps, P., Bowern, C., Hansen, C., Hill, J. & Zentz, J. On numeral complexity in hunter-gatherer languages. *Linguistic Typology* **16**, 41–109 (2012).

29 Butterworth, B., Reeve, R., Reynolds, F. & Lloyd, D. Numerical thought with and without words: Evidence from indigenous Australian children. *Proceedings of the National Academy of Sciences of the USA* **105**, 13179–13184 (2008).

30 Raghubar, K.P., Barnes, M.A. & Hecht, S.A. Working memory and mathematics: A review of developmental, individual difference, and cognitive approaches. *Learning and Individual Differences* **20**, 110–122 (2010).

31 Reeve, R., Reynolds, F., Paul, J. & Butterworth, B. Culture-Independent prerequisites for early arithmetic. *Psychological Science* **29**, 1383–1392 (2018).

32 Kearins, J. Visual spatial memory of Australian Aboriginal children of desert regions. *Cognitive Psychology* **13**, 434–460 (1981).

33 Diamond, J. *Guns, Germs and Steel: The Fates of Human Societies* (Jonathan Cape, 1997).

34 Cantlon, J., Fink, R., Safford, K. & Brannon, E.M. Heterogeneity impairs numerical matching but not numerical ordering in preschool children. *Developmental Science* **10**, 431–440 (2007).

35 Wynn, K. Addition and subtraction by human infants. *Nature* **358**, 749–751 (1992).

36 McCrink, K. & Wynn, K. Large-Number addition and subtraction by 9-month-old infants. *Psychological Science* **15**, 776–781 (2004).

37 Jordan, K.E. & Brannon, E.M. The multisensory representation of number in infancy. *Proceedings of the National Academy of Sciences of the United States of America* **103**, 3486–3489 (2006).

38 Izard, V., Sann, C., Spelke, E.S. & Streri, A. Newborn infants perceive abstract numbers. *Proceedings of the National Academy of Sciences* **106**, 10382–10385 (2009).

39 Burr, D.C. & Ross, J. A visual sense of number. *Current Biology* **18**, 425–428 (2008).

40 Whalen, J., Gallistel, C.R. & Gelman, R. Nonverbal counting in humans: The Psychophysics of number representation. *Psychological Science* **10**, 130–137 (1999).

41 Cordes, S., Gelman, R., Gallistel, C.R. & Whalen, J. Variability signatures distinguish verbal from nonverbal counting for both large and small numbers. *Psychonomic Bulletin & Review* **8**, 698–707 (2001).

42 Hartnett, P. & Gelman, R. Early understandings of numbers: Paths or barriers to the construction of new understandings? *Learning and Instruction* **8**, 341–374 (1998).

43 Cheung, P., Rubenson, M. & Barner, D. To infinity and beyond: Children generalize the successor function to all possible numbers years after learning to count. *Cognitive Psychology* **92**, 22–36 (2017).

44 Sarnecka, B.W. & Gelman, S.A. Six does not just mean a lot: preschoolers see number words as specific. *Cognition* **92**, 329–352 (2004).

45 Cipolotti, L., Butterworth, B. & Denes, G. A specific deficit for numbers in a case of dense acalculia. *Brain* **114**, 2619–2637 (1991).

46 Warrington, E.K. & James, M. Tachistoscopic number estimation in patients with unilateral lesions. *Journal of Neurology, Neurosurgery and Psychiatry* **30**, 468–474 (1967).

47 Vetter, P., Butterworth, B. & Bahrami, B. A candidate for the attentional bottleneck: Set-size Specific modulation of the right TPJ during attentive enumeration. *Journal of Cognitive Neuroscience* **23**, 728–736 (2010).

48 Arsalidou, M. & Taylor, M.J. Is 2+2=4? Meta-analyses of brain areas needed for numbers and calculations. *NeuroImage* **54**, 2382–2393 (2011). Piazza, M., Mechelli, A., Butterworth, B. & Price, C.J. Are subitizing and counting implemented as separate or functionally overlapping processes? *NeuroImage* **15**, 435–446 (2002).

49 Castelli, F., Glaser, D.E. & Butterworth, B. Discrete and analogue quantity processing in the parietal lobe: A functional MRI study. *Proceedings of the National Academy of Sciences of the United States of America* **103**, 4693–4698 (2006).

50 Piazza, M., Mechelli, A., Price, C.J. & Butterworth, B. Exact and approximate judgements of visual and auditory numerosity: An fMRI study. *Brain Research* **1106**, 177–188 (2006).

51 Santens, S., Roggeman, C., Fias, W. & Verguts, T. Number processing pathways in human parietal cortex. *Cerebral Cortex* **20**, 77–88 (2010).

52 Line drawings from https://neupsykey.com/2-landmarks/

53 Pesenti, M. et al. Mental calculation expertise in a prodigy is sustained by right prefrontal and medial-temporal areas. *Nature Neuroscience* **4**, 103–107 (2001). Butterworth, B. What makes a prodigy? *Nature Neuroscience* **4**, 11–12 (2001).

54 Aydin, K. et al. Increased gray matter density in the parietal cortex of mathematicians: A Voxel-Based Morphometry study. *American Journal of Neuroradiology* **28**, 1859–1864 (2007). Amalric, M. & Dehaene, S. Origins of the brain networks for advanced mathematics in expert mathematicians. *Proceedings of the National Academy of Sciences* **113**, 4909–4917 (2016).

55 Alarcon, M., Defries, J., Gillis Light, J. & Pennington, B. A twin study of mathematics disability. *Journal of Learning Disabilities* **30**, 617–623 (1997).

56 Kovas, Y., Haworth, C.M., Dale, P.S. & Plomin, R. The genetic and environmental origins of learning abilities and disabilities in the early school years. *Monograph of the Society for Research in Child Development* **72**, 1–144 (2007).

57 Tosto, M.G. et al. Why do we differ in number sense? Evidence from a genetically sensitive investigation. *Intelligence* **43**, 35–46 (2014).

58 Bishop, D.V.M. & Snowling, M. Developmental dyslexia and specific language impairment: Same or different? *Psychological Bulletin* **130**, 858–886 (2004). Paulesu, E. et al. Dyslexia: Cultural diversity and biological unity. *Science* **291**, 2165 (2001). Stein, J. & Walsh, V. To see but not to read: The magnocellular theory of dyslexia. *Trends in Neurosciences* **20**, 147–152 (1997). Zorzi, M. et al. Extra-large letter spacing improves reading in dyslexia. *Proceedings of the National Academy of Sciences* **109**, 11455–11459 (2012).

59 Butterworth, B. *Dyscalculia: From Science to Education* (Routledge, 2019). Butterworth, B., Varma, S. & Laurillard, D. Dyscalculia: From brain to education. *Science* **332**, 1049–1053 (2011). Piazza, M. et al. Developmental trajectory of number acuity reveals a severe impairment in developmental dyscalculia. *Cognition* **116**, 33–41 (2010).

60 Ranpura et al, under review.

61 Gautam, P., Nuñez, S.C., Narr, K.L., Kan, E.C. & Sowell, E.R. Effects of prenatal alcohol exposure on the development of white matter volume and change in executive function. *NeuroImage: Clinical* **5**, 19–27 (2014). Kopera-Frye, K., Dehaene, S. & Streissguth, A.P. Impairments of number processing induced by prenatal alcohol exposure. *Neuropsychologia* **34**, 1187–1196 (1996).

62 Isaacs, E.B., Edmonds, C.J., Lucas, A. & Gadian, D.G. Calculation difficulties in children of very low birthweight: A neural correlate. *Brain* **124**, 1701–1707 (2001).

63 Butterworth, B. et al. Language and the origins of number skills: Karyotypic differences in Turner's syndrome. *Brain & Language* **69**, 486–488 (1999). Bruandet, M., Molko, N., Cohen, L. & Dehaene, S. A cognitive characterization of dyscalculia in Turner syndrome. *Neuropsychologia* **42**, 288–298 (2004). Molko, N. et al. Functional and structural alterations of the intraparietal sulcus in a developmental dyscalculia of genetic origin. *Neuron* **40**, 847–858 (2003).

64 Semenza, C. et al. Genetics and mathematics: FMR1 premutation female carriers. *Neuropsychologia* **50**, 3757–3763 (2012).

65 Baron-Cohen, S. et al. A genome wide association study of mathematical ability reveals an association at chromosome 3q29, a locus associated with autism and learning difficulties: A preliminary study. *PLOS ONE* **9**, e96374, doi:10.1371/journal.pone.0096374 (2014). Pettigrew, K.A. et al. Lack of replication for the myosin-18B association with mathematical ability in independent cohorts. *Genes, Brain and Behavior* **14**, 369–376 (2015).

Chapter 3

1 Friberg, J. Numbers and measures in the earliest written records. *Scientific American* **250**, 78–85 (1984).

2 Friberg, J. Three thousand years of sexagesimal numbers in Mesopotamian mathematical texts. *Archive for History of Exact Sciences* **73**, 183–216 (2019).

3 Mattessich, R. Recent insights into Mesopotamian accounting of the 3rd millennium BCE – successor to token accounting. *Accounting Historians Journal* **25**, 1–27 (1998).

4 Ifrah, G. *The Universal History of Numbers. From Prehistory to the Invention of the Computer* (Harvill Press, 1998).

5   Vega, G. *The Royal Commentaries of the Incas (Comentarios reales de los Incas)* (1609; Ediciones el Lector, 2008).

6   Ascher, M. & Ascher, R. *Code of the Quipu: A Study in Media, Mathematics, and Culture* (University of Michigan Press, 1981).

7   Hyland, S., Ware, G.A. & Clarke, M. Knot direction in a khipu/ alphabetic text from the Central Andes. *Latin American Antiquity* **25**, 189–197 (2014).

8   Hyland, S. Ply, Markedness, and redundancy: New evidence for how Andean khipus encoded information. *American Anthropologist* **116**, 643–648 (2014).

9   Quilter, J. et al. Traces of a lost language and number system discovered on the north coast of Peru. *American Anthropologist* **112**, 357–369 (2010).

10 https://mayaarchaeologist.co.uk/2016/12/28/maya-numbers/

11 Sharer, R.J. *The Ancient Maya. Fifth Edition* (Stanford University Press, 1994).

12 D'Errico, F. et al. From number sense to number symbols: An archaeological perspective. *Philosophical Transactions of the Royal Society B: Biological Sciences* **373** (2018).

13 Flegg, G. (Macmillan in association with the Open University, London, 1989).

14 Powell, A., Shennan, S. & Thomas, M.G. Late Pleistocene demography and the appearance of modern human behavior. *Science* **324**, 1298 (2009).

15 D'Errico, F. et al. From number sense to number symbols. An archaeological perspective. *Philosophical Transactions of the Royal Society B: Biological Sciences* **373** (2018).

16 D'Errico, F. Technology, motion and the meaning of epipalaeolithic art. *Current Anthropology* **33**, 94–109 (1992).

17 Henshilwood, C.S., d'Errico, F. & Watts, I. Engraved ochres from the Middle Stone Age levels at Blombos Cave, South Africa. *Journal of Human Evolution* **57**, 27–47 (2009).

18 D'Errico, F. et al. The technology of the earliest European cave paintings: El Castillo Cave, Spain. *Journal of Archaeological Science* **70**, 48–65 (2016).

19 Chauvet, J.-M., Deschamps, E.B. & Hillaire, C. *Chauvet Cave: The Discovery of the World's Oldest Paintings* (Thames & Hudson, 1996).

20 Clottes, J. *Les Cavernes de Niaux* (Editions du Seuil, 1995).

21 https://www.youtube.com/watch?v=R1R8yrEGAgw

22 Hoffmann, D.L. et al. U-Th dating of carbonate crusts reveals Neanderthal origin of Iberian cave art. *Science* **359**, 91 (2018).

23 Hardy, B.L. et al. Direct evidence of Neanderthal fibre technology and its cognitive and behavioral implications. *Scientific Reports* **10**, 4889 (2020).

24 Joordens, J.C.A. et al. *Homo erectus* at Trinil on Java used shells for tool production and engraving. *Nature* **518**, 228–231 (2015).

25 Lewis-Williams, D. *Conceiving God: The Cognitive Origin and Evolution of Religion* (Thames & Hudson, 2011).

26 Pagel, M. & Meade, A. The deep history of the number words. *Philosophical Transactions of the Royal Society B: Biological Sciences* **373** (2018).

27 Bowern, C. & Zentz, J. Diversity in the numeral systems of Australian languages. *Anthropological Linguistics* **54**, 133–160 (2012).

28 Dixon, R.M.W. *The Languages of Australia* (Cambridge University Press, 1980).

29 Kendon, A. *Sign Languages of Aboriginal Australia: Cultural, Semiotic and Communicative Perspectives* (Cambridge University Press, 1988).

Chapter 4

1 Matsuzawa, T. The Ai project: Historical and ecological context. *Animal Cognition* **6**, 199–211, doi:10.1007/s10071-003-0199-2 (2003).

2 Fouts, R. & Mills, S. *Next of Kin* (Michael Joseph, 1997).

3 Tomonaga, M. & Matsuzawa, T. Enumeration of briefly presented items by the chimpanzee (*Pan troglodytes*) and humans (*Homo sapiens*). *Animal Learning and Behavior* **30**, 143–157 (2002).

4 Inoue, S. & Matsuzawa, T. Working memory of numerals in chimpanzees. *Current Biology* **17**, R1004–R1006 (2007).

5 Matsuzawa, T. in *Cognitive Development in Chimpanzees* (eds T. Matsuzawa, M. Tomonaga & M. Tanaka), 3–33 (Springer Tokyo, 2006).

6 Menzel, E.W. Chimpanzee spatial memory organization. *Science* **182**, 943, doi:10.1126/science.182.4115.943 (1973).

7 Boesch, C. Teaching among wild chimpanzees. *Animal Behaviour* **41**, 530–532 (1991).

8 Biro, D., Sousa, C. & Matsuzawa, T. in *Cognitive Development in Chimpanzees* (eds T. Matsuzawa, M. Tomonaga & M. Tanaka) 476–508 (Springer, 2006).

9 Hanus, D. & Call, J. Discrete quantity judgments in the great apes (*Pan paniscus, Pan troglodytes, Gorilla gorilla, Pongo pygmaeus*): The effect of presenting whole sets versus item-by-item. *Journal of Comparative Psychology* **121**, 241–249 (2007).

10 Martin, C.F., Biro, D. & Matsuzawa, T. Chimpanzees spontaneously take turns in a shared serial ordering task. *Scientific Reports* **7**, 14307, doi:10.1038/s41598-017-14393-x (2017).

11 Boesch, C. Symbolic communication in wild chimpanzees? *Human Evolution* **6**, 81–89, doi:10.1007/BF02435610 (1991).

12 Wilson, M.L., Hauser, M.D. & Wrangham, R.W. Does partici-
    pation in intergroup conflict depend on numerical assessment,
    range location, or rank for wild chimpanzees? *Animal Behavior*
    61, 1203–1216 (2001).

13 Boysen, S.T. in *The Development of Numerical Competence: Animal
    and Human Models Comparative Cognition and Neuroscience* (eds
    S.T. Boysen & E.J. Capaldi) (LEA, 1993).

14 Brannon, E.M. & Terrace, H.S. Ordering of the numerosities 1 to
    9 by monkeys. *Science* **282**, 746–749 (1998).

15 Sawamura, H., Shima, K. & Tanji, J. Numerical representation for
    action in the parietal cortex of the monkey. *Nature* **415**, 918–922 (2002).

16 Davis, H. & Pérusse, R. Numerical competence in animals: Def-
    initional issues, current evidence and a new research agenda.
    Behavioral and Brain Sciences **11**, 561–579 (1988).

17 Davis, H. & Memmott, J. Counting behavior in animals: A critical
    evaluation. *Psychological Bulletin* **92**, 547–571, https://doi.org/10.
    1037/0033-2909.92.3.547 (1982).

18 Cantlon, J.F. & Brannon, E.M. How much does number matter to
    a monkey (*Macaca mulatta*)? *Journal of Experimental Psychology:
    Animal Behavior Processes* **33**, 32–41, https://doi.org/10.1037/0097-
    7403.33.1.32 (2007).

19 Cantlon, J.F. & Brannon, E.M. Shared system for ordering small
    and large numbers in monkeys and humans. *Psychological Science*
    17, 401–406, doi:10.1111/j.1467-9280.2006.01719.x (2006).

20 Cantlon, J.F. & Brannon, E.M. Basic math in monkeys and college
    students. *PLOS Biology* **5**, e328, doi:10.1371/journal.pbio.0050328
    (2007).

21 Livingstone, M.S. et al. Symbol addition by monkeys provides evi-
    dence for normalized quantity coding. *Proceedings of the National
    Academy of Sciences* **111**, 6822, doi:10.1073/pnas.1404208111 (2014).

22 Hauser, M.D., Carey, S. & Hauser, L.B. Spontaneous number representation in semi-free-ranging rhesus monkeys. *Proceedings of the Royal Society B* **267**, 829–833 (2000).

23 Jordan, K.E., Brannon, E.M., Logothetis, N.K. & Ghazanfar, A.A. Monkeys match the number of voices they hear to the number of faces they see. *Current Biology* **15**, 1034–1038 (2005).

24 Jordan, K.E. & Brannon, E.M. The multisensory representation of number in infancy. *Proceedings of the National Academy of Sciences of the United States of America* **103**, 3486–3489 (2006).

25 Flombaum, J.I., Jungea, J.A. & Hauser, M.D. Rhesus monkeys (*Macaca mulatta*) spontaneously compute addition operations over large numbers. *Cognition* **97**, 315–325 (2005).

26 Brotcorne, F. et al. Intergroup variation in robbing and bartering by long-tailed macaques at Uluwatu Temple (Bali, Indonesia). *Primates* **58**, 505–516, doi:10.1007/s10329-017-0611-1 (2017).

27 Leca, J.-B., Gunst, N., Gardiner, M. & Nengah Wandia, I. Acquisition of object-robbing and object/food-bartering behaviours: A culturally maintained token economy in freeranging long-tailed macaques. *Philosophical Transactions of the Royal Society B* **376:20190677**, https://doi.org/10.1098/rstb.2019.0677 (2021).

28 Ratcliffe, R. Bali's thieving monkeys can spot high-value items to ransom. *Guardian* (2021).

29 Cantlon, J.F., Piantadosi, S.T., Ferrigno, S., Hughes, K.D. & Barnard, A.M. The origins of counting algorithms. *Psychological Science* **26**, 853–865, doi:10.1177/0956797615572907 (2015).

30 Strandburg-Peshkin, A., Farine, D.R., Couzin, I.D. & Crofoot, M.C. Shared decision-making drives collective movement in wild baboons. *Science* **348**, 1358, doi:10.1126/science.aaa5099 (2015).

31 Santens, S., Roggeman, C., Fias, W. & Verguts, T. Number pro-
    cessing pathways in human parietal cortex. *Cerebral Cortex* **20**,
    77–88, doi:10.1093/cercor/bhp080 (2010).

32 Castelli, F., Glaser, D.E. & Butterworth, B. Discrete and analogue
    quantity processing in the parietal lobe: A functional MRI study.
    *Proceedings of the National Academy of Sciences of the United States of
    America* **103**, 4693–4698 (2006).

33 Piazza, M., Mechelli, A., Price, C.J. & Butterworth, B. Exact and
    approximate judgements of visual and auditory numerosity: An
    fMRI study. *Brain Research* **1106**, 177–188 (2006).

34 Roitman, J.D., Brannon, E.M. & Platt, M.L. Monotonic coding
    of numerosity in macaque lateral intraparietal area. *PLOS Biology*,
    doi:10.1371/journal.pbio.0050208 (2007).

35 Della Puppa, A. et al. Right parietal cortex and calculation pro-
    cessing: Intraoperative functional mapping of multiplication and
    addition in patients affected by a brain tumor. *Journal of Neuro-
    surgery* **119**, 1107–1111, doi:10.3171/2013.6.JNS122445 (2013).

36 Nieder, A., Freedman, D.J. & Miller, E.K. Representation of the
    quantity of visual items in the primate prefrontal cortex. *Science*
    297, 1708–1711 (2002).

37 Nieder, A., Diester, I. & Tudusciuc, O. Temporal and spatial enu-
    meration processes in the primate parietal cortex. *Science* **313**,
    1431–1435 (2006).

38 Nieder, A. *A Brain for Numbers: The Biology of the Number Instinct*
    (MIT Press, 2019).

39 Semenza, C., Salillas, E., De Pallegrin, S. & Della Puppa, A. Bal-
    ancing the two Hemispheres in simple calculation: Evidence from
    direct cortical electrostimulation. *Cerebral Cortex* **27**, 4806–4814,
    doi:10.1093/cercor/bhw277 (2017).

40 Salillas, E. et al. A MEG study on the processing of time and quantity: Parietal overlap but functional divergence. *Frontiers in Psychology* **10**, 139 (2019).

41 Zhao, H. et al. Arithmetic learning modifies the functional connectivity of the fronto-parietal network. *Cortex* **111**, 51–62, https://doi.org/10.1016/j.cortex.2018.07.016 (2019).

42 Matejko, A.A. & Ansari, D. Drawing connections between white matter and numerical and mathematical cognition: A literature review. *Neuroscience & Biobehavioral Reviews* **48**, 35–52, http://dx.doi.org/10.1016/j.neubiorev.2014.11.006 (2015).

Chapter 5

1 Grinnell, J., Packer, C. & Pusey, A.E. Cooperation in male lions: Kinship, reciprocity or mutualism? *Animal Behaviour* **49**, 95–105 (1995).

2 McComb, K., Packer, C. & Pusey, A. Roaring and numerical assessment in contests between groups of female lions, *Panthera leo. Animal Behaviour* **47**, 379–387 (1994).

3 McComb, K. Female choice for high roaring rates in red deer, *Cervus elaphus. Animal Behaviour* **41**, 79–88 (1991).

4 Benson-Amram, S., Gilfillan, G. & McComb, K. Numerical assessment in the wild: Insights from social carnivores. *Philosophical Transactions of the Royal Society B: Biological Sciences* **373** (2018).

5 Benson-Amram, S., Heinen, V.K., Dryer, S.L. & Holekamp, K.E. Numerical assessment and individual call discrimination by wild spotted hyaenas, *Crocuta crocuta. Animal Behaviour* **82**, 743–752 (2011).

6   Mechner, F. Probability relations within response sequences under ratio reinforcement. *Journal of the Experimental Analysis of Behavior* **1**, 109–122 (1958).

7   Meck, W.H. & Church, R.M. A mode control model of counting and timing processes. *Journal of Experimental Psychology: Animal Behavior Processes* **9**, 320–334 (1983).

8   Panteleeva, S., Reznikova, Z. & Vygonyailova, O. Quantity judgments in the context of risk/reward decision making in striped field mice: first 'count', then hunt. *Frontiers in Psychology* **4**, 53 (2013).

9   Çavdaroğlu, B. & Balcı, F. Mice can count and optimize count-based decisions. *Psychonomic Bulletin & Review* **23**, 1–6 (2015).

10 Mortensen, H.S. et al. Quantitative relationships in delphinid neocortex. *Front Neuroanat* **8**, 132 (2014).

11 Fields, R.D. Of whales and men. *Scientific American*, https://blogs.scientificamerican.com/news-blog/are-whales-smarter-than-we-are/ (2008).

12 Fox, K.C.R., Muthukrishna, M. & Shultz, S. The social and cultural roots of whale and dolphin brains. *Nature Ecology & Evolution* **1**, 1699–1705 (2017).

13 Pryor, K. & Lindbergh, J. A dolphin–human fishing cooperative in Brazil. *Marine Mammal Science* **6**, 77–82 (1990).

14 Garrigue, C., Clapham, P.J., Geyer, Y., Kennedy, A.S. & Zerbini, A.N. Satellite tracking reveals novel migratory patterns and the importance of seamounts for endangered South Pacific humpback whales. *Royal Society Open Science* **2**, 150489 (2015).

15 Patzke, N. et al. In contrast to many other mammals, cetaceans have relatively small hippocampi that appear to lack adult neurogenesis. *Brain Structure and Function* **220**, 361–383 (2015).

16 Abramson, J.Z., Hernández-Lloreda, V., Call, J. & Colmenare, F. Relative quantity judgments in the beluga whale (*Delphinapterus*

leucas) and the bottlenose dolphin (*Tursiops truncatus*). *Behavioural Processes* **96**, 11–19.

17 Kilian, A., Yaman, S., Von Fersen, L. & Güntürkün, O. A bottlenose dolphin discriminates visual stimuli differing in numerosity. *Learning and Behavior* **31**, 133–142 (2003).

18 Davis, H. & Bradford, S. A. Counting behavior by rats in a simulated natural environment. *Ethology* **73**, 265–280 (1986).

19 Suzuki, K., & Kobayashi, T. Numerical competence in rats (*Rattus norvegicus*): Davis and Bradford (1986) extended. *Journal of Comparative Psychology* **114**, 73–85 (2000).

20 Thompson, R.F., Mayers, K.S., Robertson, R.T. & Patterson, C.J. Number Coding in association cortex of the cat. *Science* **168**, 271–273 (1970).

Chapter 6

1 Pepperberg, I.M., Willner, M.R. & Gravitz, L.B. Development of Piagetian object permanence in grey parrot (*Psittacus erithacus*). *Journal of Comparative Psychology* **111**, 63–75 (1997).

2 Pepperberg, I.M. Acquisition of the same/different concept by an African Grey parrot (*Psittacus erithacus*): Learning with respect to categories of color, shape, and material. *Animal Learning & Behavior* **15**, 423–432 (1987).

3 Pepperberg, I.M. Numerical competence in an African gray parrot (*Psittacus erithacus*). *Journal of Comparative Psychology* **108**, 36–44 (1994).

4 Pepperberg, I.M. Grey parrot (*Psittacus erithacus*) numerical abilities: Addition and further experiments on a zero-like concept. *Journal of Comparative Psychology* **120**, 1–11 (2006).

5 Pepperberg, I.M. & Carey, S. Grey parrot number acquisition: The inference of cardinal value from ordinal position on the numeral list. *Cognition* **125**, 219–232 (2012).

6 Sarnecka, B.W. & Gelman, S.A. Six does not just mean a lot: preschoolers see number words as specific. *Cognition* **92**, 329–352 (2004).

7 Pepperberg, I.M. in *Mathematical Cognition and Learning*, Vol. 1 (eds D.C. Geary, D.B. Berch & K. Mann Koepke), 67–89 (Elsevier, 2015).

8 Koehler, O. The ability of birds to count. *Bulletin of Animal Behaviour* **9**, 41–45 (1950).

9 Koehler, O. „Zähl"-Versuche an einem Kolkraben und Vergleichsversuche an Menschen. *Zeitschrift für Tierpsychologie* **5**, 575–712 (1943).

10 Thorpe, W.H. *Learning and Instinct in Animals*, second edn (Methuen, 1963).

11 Ditz, H.M. & Nieder, A. Neurons selective to the number of visual items in the corvid songbird endbrain. *Proceedings of the National Academy of Sciences* **112**, 7827–7832 (2015).

12 Scarf, D., Hayne, H. & Colombo, M. Pigeons on par with primates in numerical competence. *Science* **334**, 1664 (2011).

13 Rugani, R. Towards numerical cognition's origin: Insights from day-old domestic chicks. *Philosophical Transactions of the Royal Society B: Biological Sciences* **373**, 2016.0509 (2018).

14 Rugani, R., Fontanari, L., Simoni, E., Regolin, L. & Vallortigara, G. Arithmetic in newborn chicks. *Proceedings of the Royal Society B* **276**, 2451–2460 (2009).

15 Rilling, M. in *The Development of Numerical Competence: Animal and Human Models Comparative Cognition and Neuroscience* (eds S.T. Boysen & E.J. Capaldi) (LEA, 1993).

16 Lyon, B. Egg recognition and counting reduce costs of avian con-specific brood parasitism. *Nature* **422**, 495–499 (2003).

17 White, D.J., Ho, L. & Freed-Brown, G. Counting chicks before they hatch: Female cowbirds can time readiness of a host nest for parasitism. *Psychological Science* **20**, 1140–1145 (2009).

18 Searcy, W.A. & Nowicki, S. Birdsong learning, avian cognition and the evolution of language. *Animal Behaviour* **151**, 217–227 (2019).

19 Nottebohm, F. The neural basis of birdsong. *PLOS Biology* **3**, e164 (2005).

20 Gill, R.E. et al. Hemispheric-scale wind selection facilitates bar-tailed godwit circum-migration of the Pacific. *Animal Behaviour* **90**, 117130 (2014).

21 Åkesson, S. & Bianco, G. Assessing vector navigation in long-distance migrating birds. *Behavioral Ecology* **27**, 865–875 (2016).

22 Armstrong, C. et al. Homing pigeons respond to time-compensated solar cues even in sight of the loft. *PLOS ONE* **8**, e63130 (2013).

23 Padget, O. et al. Shearwaters know the direction and distance home but fail to encode intervening obstacles after free-ranging foraging trips. *Proceedings of the National Academy of Sciences* **116**, 21629 (2019).

24 Thorup, K. et al. Evidence for a navigational map stretching across the continental US in a migratory songbird. *Proceedings of the National Academy of Sciences* **104**, 18115 (2007).

25 https://sites.google.com/site/michaelhammondhistoryofscience/project/chip-log

26 Collett, T.S. Path integration: how details of the honeybee waggle dance and the foraging strategies of desert ants might help in understanding its mechanisms. *Journal of Experimental Biology* **222**, jeb205187 (2019).

27 Gallistel, C.R. Finding numbers in the brain. *Philosophical Transactions of the Royal Society B: Biological Sciences* **373**, 2017.0119 (2018).

28 Gallistel, C.R. in *The Sailing Mind: Studies in Brain and Mind* (ed. Roberto Casati) (forthcoming).

29 Olkowicz, S. et al. Birds have primate-like numbers of neurons in the forebrain. *Proceedings of the National Academy of Sciences* **113**, 7255–7260 (2016).

30 O'Keefe, J. & Dostrovsky, J. The hippocampus as a spatial map: Preliminary evidence from unit activity in the freely-moving rat. *Brain Research* **34**, 171–175 (1971).

31 Wirthlin, M. et al. Parrot genomes and the evolution of heightened longevity and cognition. *Current Biology* **28**, 4001–4008.e7 (2018).

32 Lovell, P.V., Huizinga, N.A., Friedrich, S.R., Wirthlin, M. & Mello, C.V. The constitutive differential transcriptome of a brain circuit for vocal learning. *BMC Genomics* **19**, 231 (2018).

Chapter 7

1 Naumann, R. et al. The reptilian brain. *Current Biology* **25**, R317–R321 (2015).

2 Northcutt, R.G. Understanding vertebrate brain evolution. *Integrative and Comparative Biology* **42**, 743–756 (2002).

3 Uller, C., Jaeger, R., Guidry, G. & Martin, C. Salamanders (*Plethodon cinereus*) go for more: Rudiments of number in an amphibian. *Animal Cognition* **6**, 105–112 (2003).

4 Hauser, M.D., Carey, S. & Hauser, L.B. Spontaneous number representation in semi-free-ranging rhesus monkeys. *Proceedings of the Royal Society B* **267**, 829–833 (2000).

5 Miletto Petrazzini, M.E. et al. Quantitative abilities in a reptile (*Podarcis sicula*). *Biology Letters* **13**, 2016.0899 (2017).

6 Klump, G.M. & Gerhardt, H.C. Use of non-arbitrary acoustic criteria in mate choice by female gray tree frogs. *Nature* **326**, 286–288 (1987).

7 Rose, G.J. The numerical abilities of anurans and their neural correlates: Insights from neuroethological studies of acoustic communication. *Philosophical Transactions of the Royal Society B: Biological Sciences* **373** (2018).

8 Angier, N. in *New York Times* (2018).

9 Gerhardt, H.C., Roberts, J.D., Bee, M.A. & Schwartz, J.J. Call matching in the quacking frog (*Crinia georgiana*). *Behavioral Ecology and Sociobiology* **48**, 243–251 (2000).

10 Balestrieri, A., Gazzola, A., Pellitteri-Rosa, D. & Vallortigara, G. Discrimination of group numerousness under predation risk in anuran tadpoles. *Animal Cognition* **22**, 223–230 (2019).

11 Stancher, G., Rugani, R., Regolin, L. & Vallortigara, G. Numerical discrimination by frogs (*Bombina orientalis*). *Animal Cognition* **18**, 219–229 (2015).

12 MacLean, P.D. *The Triune Brain in Evolution: Role in Paleocerebral Functions* (Plenum Press, 1990).

13 Davis, H. & Pérusse, R. Numerical competence in animals: Definitional issues, current evidence and a new research agenda. *Behavioral and Brain Sciences* **11**, 561–579 (1988).

14 Miletto Petrazzini, M.E., Bertolucci, C. & Foà, A. Quantity discrimination in trained lizards (*Podarcis sicula*). *Frontiers in Psychology* **9**, 274 (2018).

15 Gazzola, A., Vallortigara, G. & Pellitteri-Rosa, D. Continuous and discrete quantity discrimination in tortoises. *Biology Letters* **14**, 2018.0649 (2018).

16 Darwin, C. Perception in the lower animals. *Nature* **7**, 360 (1873).

17 Gould, James L. Animal navigation: Memories of home. *Current Biology* **25**, R104–R106 (2015).

18 Brothers, J.R. & Lohmann, Kenneth J. Evidence for geomagnetic imprinting and magnetic navigation in the natal homing of sea turtles. *Current Biology* **25**, 392–396 (2015).

Chapter 8

1 Agrillo, C. & Bisazza, A. Understanding the origin of number sense: A review of fish studies. *Philosophical Transactions of the Royal Society B: Biological Sciences* **373** (2018).

2 Thorpe, W.H. *Learning and Instinct in Animals*, 2nd ed. (Methuen, 1963).

3 Tinbergen, N. The Curious behavior of the stickleback. *Scientific American* **187**, 22–27 (1952).

4 Agrillo, C., Dadda, M., Serena, G. & Bisazza, A. Do fish count? Spontaneous discrimination of quantity in female mosquitofish. *Animal Cognition* **11**, 495–503 (2008).

5 Hager, M.C. & Helfman, G.S. Safety in numbers: shoal size choice by minnows under predatory threat. *Behavioral Ecology and Sociobiology* **29**, 271–276 (1991).

6 Frommen, J.G., Hiermes, M. & Bakker, T.C.M. Disentangling the effects of group size and density on shoaling decisions of three-spined sticklebacks (*Gasterosteus aculeatus*). *Behavioral Ecology and Sociobiology* **63**, 1141–1148 (2009).

7 Agrillo, C., Piffer, L., Bisazza, A. & Butterworth, B. Evidence for two numerical systems that are similar in humans and guppies. *PLOS ONE* **7**, e31923, doi:10.1371/journal.pone.0031923 (2012).

8 Vetter, P., Butterworth, B. & Bahrami, B. A candidate for the attentional bottleneck: Set-size specific modulation of the right

TPJ during attentive enumeration. *Journal of Cognitive Neuroscience* **23**, 728–736, doi:10.1162/jocn.2010.21472 (2010).

9 Dadda, M., Piffer, L., Agrillo, C. & Bisazza, A. Spontaneous number representation in mosquitofish. *Cognition* **112**, 343–348 (2009).

10 Bisazza, A. et al. Collective enhancement of numerical acuity by meritocratic leadership in fish. *Scientific Reports* **4**, doi:10.1038/srep04560 (2014).

11 Miletto Petrazzini, M.E., Agrillo, C., Izard, V. & Bisazza, A. Relative versus absolute numerical representation in fish: Can guppies represent 'fourness'? *Animal Cognition* **18**, 1007–1017 (2015).

12 Agrillo, C., Dadda, M., Serena, G. & Bisazza, A. Use of number by fish. *PLOS ONE* **4**, doi:10.1371/journal.pone.0004786 (2009).

13 Bahrami, B., Didino, D., Frith, C., Butterworth, B. & Rees, G. Collective enumeration. *Journal of Experimental Psychology: Human Perception and Performance* **39**, 338–347, doi:10.1037/a0029717 (2013).

14 Butterworth, B. *Dyscalculia: From Science to Education* (Routledge, 2019).

15 Ward, A.J.W. et al. Initiators, leaders, and recruitment mechanisms in the collective movements of damselfish. *The American Naturalist* **181**, 748–760, doi:10.1086/670242 (2013).

16 Glasauer, S.M.K. & Neuhauss, S.C.F. Whole-genome duplication in teleost fishes and its evolutionary consequences. *Molecular Genetics and Genomics* **289**, 1045–1060, doi:10.1007/s00438-014-0889-2 (2014).

17 Wang, S. et al. Evolutionary and expression analyses show co-option of khdrbs genes for origin of vertebrate brain. *Frontiers in Genetics* **8**, 225 (2018).

18 Messina, A., Potrich, D., Schiona, I., Sovrano, V.A., Fraser, S.E., Brennan, C.H., et al. Neurons in the Dorso-Central Division of Zebrafish Pallium Respond to Change in Visual Numerosity. *Cerebral Cortex* https://doi.org/10.1093/cercor/bhab218 (2021).

19 Thorpe, W.H. *Learning and Instinct in Animals* (Methuen, 1963).

Chapter 9

1 Polilov, A.A. & Makarova, A.A. The scaling and allometry of organ size associated with miniaturization in insects: A case study for *Coleoptera* and *Hymenoptera*. *Scientific Reports* 7 (2017).

2 Eberhard, W.G. & Wcislo, W.T. Plenty of room at the bottom. *American Scientist* 100, 226–233 (2012).

3 von Frisch, K. *The Dance Language and Orientation of Bees* (Harvard University Press, 1967).

4 Papi, F. Animal navigation at the end of the century: A retrospect and a look forward. *Italian Journal of Zoology* 68, 171–180 (2001).

5 Stone, T. et al. An anatomically constrained model for path integration in the bee brain. *Current Biology* 27, 3069–3085.e3011 (2017).

6 Skorupski, P., MaBouDi, H., Galpayage Dona, H.S. & Chittka, L. Counting insects. *Philosophical Transactions of the Royal Society B: Biological Sciences* 373 (2018).

7 Chittka, L., Geiger, K. Can honey bees count landmarks? *Animal Behaviour* 49, 159–64 (1995).

8 Dacke, M., Srinivasan, M. Evidence for counting in insects. *Animal Cognition*, 2008;11(7):683–9.

9 Collett, T.S. Path integration: How details of the honeybee waggle dance and the foraging strategies of desert ants might help in

understanding its mechanisms. *Journal of Experimental Biology* **222**, jeb205187 (2019).

10   Couvillon, M.J., Schürch, R. & Ratnieks, F.L.W. Waggle dance distances as integrative indicators of seasonal foraging challenges. *PLOS ONE* (2014).

11   Seid, M., Seid, M.A., Castillo, A. & Wcislo, W.T. The allometry of brain miniaturization in ants. *Brain, Behavior and Evolution* **77**, 5–13 (2011).

12. Papi, F. Animal navigation at the end of the century: A retrospect and a look forward. *Italian Journal of Zoology* **68**, 171–180 (2001).

13   Huber, R. & Knaden, M. Egocentric and geocentric navigation during extremely long foraging paths of desert ants. *Journal of Comparative Physiology A* **201**, 609–616 (2015).

14   Wittlinger, M., Wehner, R. & Wolf, H. The ant odometer: Stepping on stilts and stumps. *Science* **312**, 1965 (2006).

15   Wittlinger, M., Wehner, R. & Wolf, H. The desert ant odometer: A stride integrator that accounts for stride length and walking speed. *Journal of Experimental Biology* **210**, 198 (2007).

16   D'Ettorre, P., Meunier, P., Simonelli, P. & Call, J. Quantitative cognition in carpenter ants. *Behavioral Ecology and Sociobiology* **75**, 86 (2021).

17   Cammaerts, M.-C., Cammaerts, R. Influence of Shape, Color, Size and Relative Position of Elements on Their Counting by an Ant. International *Journal of Biology*, **12**, 13–25 (2020).

18   Gross, H. et al. Number-based visual generalisation in the honeybee. *PLOS ONE* **4**, e4263 (2009).

19   Howard, S.R., Avarguès-Weber, A., Garcia, J.E., Greentree, A.D. & Dyer, A.G. Numerical cognition in honeybees enables addition and subtraction. *Science Advances* **5**, eaav0961 (2019).

20 Bortot, M. et al. Honeybees use absolute rather than relative numerosity in number discrimination. *Biology Letters* **15**, 2019.0138 (2019).

21 Howard, S.R., Avarguès-Weber, A., Garcia, J.E., Greentree, A.D. & Dyer, A.G. Numerical ordering of zero in honey bees. *Science* **360**, 1124 (2018).

22 Bortot, M., Stancher, M. & Vallortigara, G. Transfer from number to size reveals abstract coding of magnitude in honeybees. *iScience* (2020).

23 Carazo, P., Fernández-Perea, R., Font, E. Quantity Estimation Based on Numerical Cues in the Mealworm Beetle (*Tenebrio molitor*). *Frontiers in Psychology*, 2012;3.

24 Gould, S.J. *Ever since Darwin: Reflections in Natural History*. New York: W W Norton & Co (1977).

25 Karban, R., Black, C.A. & Weinbaum, S.A. How 17-year cicadas keep track of time. *Ecology Letters* **3**, 253–256 (2000).

26 Japyassú, H.F. & Laland, K.N. Extended spider cognition. *Animal Cognition* **20**, 375–395 (2017).

27 Eberhard, W.G. & Wcislo, W.T. in *Advances in Insect Physiology* 40 (ed. J. Casas), 155–214 (Academic Press, 2011).

28 Rodríguez, R.L., Briceño, R.D., Briceño-Aguilar, E. & Höbel, G. *Nephila clavipes* spiders (Araneae: *Nephilidae*) keep track of captured prey counts: Testing for a sense of numerosity in an orb-weaver. *Animal Cognition* **18**, 307–314 (2015).

29 Davis, H. & Pérusse, R. Numerical competence in animals: Definitional issues, current evidence and a new research agenda. *Behavioral and Brain Sciences* **11**, 561–579 (1988).

30 Pollard, S.D. Robert Jackson's career understanding spider minds. *New Zealand Journal of Zoology* **43**, 4–9 (2016).

31 Cross, F.R. & Jackson, R.R. Specialised use of working memory by *Portia africana*, a spider-eating salticid. *Animal Cognition* **17**, 435–44 (2014).

32 Nelson, X.J. & Jackson, R.R. The role of numerical competence in a specialized predatory strategy of an araneophagic spider. *Animal Cognition* **15**, 699–710 (2012).

33 Vasas, V. & Chittka, L. Insect-inspired sequential inspection strategy enables an artificial network of four neurons to estimate numerosity. *iScience* **11**, 85–92 (2019).

34 MaBouDi, H. et al. Bumblebees use sequential scanning of countable items in visual patterns to solve numerosity tasks. *Integrative and Comparative Biology* **60**, 929–942 (2020).

35 Hochner, B. An Embodied View of Octopus Neurobiology. *Current Biology*, 2012;22(20):R887–R92.

36 Yang, T.-I. & Chiao, C.-C. Number sense and state-dependent valuation in cuttlefish. *Proceedings of the Royal Society B: Biological Sciences* **283**, 20161379 (2016).

37 Patel, R.N. & Cronin, T.W. Mantis shrimp navigate home using celestial and idiothetic path integration. *Current Biology* **30**, 1981–1987.e1983 (2020). Patel, R.N. & Cronin, T.W. Landmark navigation in a mantis shrimp. *Proceedings of the Royal Society B: Biological Sciences* **287**, 2020.1898 (2020).

38 Bisazza, A. & Gatto, E. Continuous versus discrete quantity discrimination in dune snail (Mollusca: *Gastropoda*) seeking thermal refuges. *Scientific Reports* **11**, 3757 (2021).

Chapter 10

1 Gallistel, C.R. Animal cognition: The representation of space, time and number. *Annual Review of Psychology* **40**, 155–189 (1989).

2 Giaquinto, M. in *The Oxford Handbook of Numerical Cognition* (eds R. Cohen Kadosh & A. Dowker), 17–31 (Oxford University Press, 2015).

3 Koehler, O. 'Zähl'-Versuche an einem Kolkraben und Vergleichsversuche an Menschen. *Zeitschrift für Tierpsychologie* **5**, 575–712 (1943).

4 Santens, S., Roggeman, C., Fias, W. & Verguts, T. Number processing pathways in human parietal cortex. *Cerebral Cortex* **20**, 77–88 (2010).

5 Roitman, J.D., Brannon, E.M. & Platt, M.L. Monotonic coding of numerosity in macaque lateral intraparietal area. *PLOS Biology*, doi:10.1371/journal.pbio.0050208 (2007).

6 Thompson, R.F., Mayers, K.S., Robertson, R.T. & Patterson, C.J. Number coding in association cortex of the cat. *Science* **168**, 271 (1970).

7 Dehaene, S. & Changeux, J.-P. Development of elementary numerical abilities: neuronal model. *Journal of Cognitive Neuroscience* **5**, 390–407 (1993).

8 Zorzi, M., Stoianov, I. & Umilta, C. in *Handbook of Mathematical Cognition* (ed. J.I.D. Campbell), 67–84 (Psychology Press, 2005).

9 Nieder, A., Freedman, D.J. & Miller, E.K. Representation of the quantity of visual items in the primate prefrontal cortex. *Science* **297**, 1708–1711 (2002).

10 Verguts, T. & Fias, W. Representation of number in animals and humans: A neural model. *Journal of Cognitive Neuroscience* **16**, 1493–1504 (2004).

11 Leslie, A.M., Gelman, R. & Gallistel, C.R. The generative basis of natural number concepts. *Trends in Cognitive Sciences* **12**, 213–218 (2008).

12　Stoianov, I., Zorzi, M. & Umiltà, C. The role of semantic and symbolic representations in arithmetic processing: Insights from simulated dyscalculia in a connectionist model. *Cortex* **40**, 194–196 (2004).

13　Butterworth, B., Varma, S. & Laurillard, D. Dyscalculia: From brain to education. *Science* **332**, 1049–1053 (2011).

14　Bisazza, A. et al. Collective enhancement of numerical acuity by meritocratic leadership in fish. *Scientific Reports* **4** (2014).

15　Tolman, E.C. Cognitive maps in rats and men. *Psychological Review* **55**, 189–208 (1948).

16　O'Keefe, J. & Nadel, L. *The Hippocampus as a Cognitive Map* (Oxford University Press, 1978).

17　Derdikman, D. & Moser, E.I. in *Space, Time and Number in the Brain* (eds S. Dehaene & E.M. Brannon), 41–57 (Academic Press, 2011).

18　Dehaene, S., Brannon, E. *Space, Time and Number in the Brain: Searching for the Foundations of Mathematical Thought* (Oxford University Press, 2011).

19　Auel, J. M. *The Clan of the Cave Bear* (Hodder & Stoughton, 1980).

20　Russell, B. *Introduction to Mathematical Philosophy* (Allen & Unwin, 1956; originally published in 1919).

21　Cordes, S., Gelman, R., Gallistel, C.R. & Whalen, J. Variability signatures distinguish verbal from nonverbal counting for both large and small numbers. *Psychonomic Bulletin & Review* **8**, 698–707 (2001).

22　Hauser, M.D., Chomsky, N. & Fitch, W.T. The faculty of language: What is it, who has it, and how did it evolve? *Science* **298**, 1569–1579 (2002).

23　Hurford, J.R. *The Linguistic Theory of Numerals* (Cambridge University Press, 1975).

24 Whitehead, A.N. *An Introduction to Mathematics* (Oxford University Press, 1948; originally published in 1911).

25 Flegg, G. *Numbers through the Ages* (Macmillan, 1989).

26 Swetz, F.J. *Capitalism and Arithmetic: The New Math of the 15th Century* (Open Court, 1987).

27 Le Guin, U.K. *The Wind's Twelve Quarters* (Harper & Row, 1975).

28 Matsuzawa, T. Use of numbers by a chimpanzee. *Nature* **315**, 57–59 (1985).

29 Cantlon, J.F. & Brannon, E.M. Basic math in monkeys and college students. *PLOS Biology* **5**, e328 (2007).

30 Mechner, F. Probability relations within response sequence maintained under ratio reinforcement. *Journal of the Experimental Analysis of Behavior* **1**, 109–121 (1958).

31 Dantzig, T. *Number: The Language of Science*, fourth ed. (Allen & Unwin, 1962).

32 Kirschhock, M.E., Ditz, H.M. & Nieder, A. Behavioral and neuronal representation of numerosity zero in the crow. *Journal of Neuroscience* **41**, 4889–4896 (2021).

33 Biro, D. & Matsuzawa, T. Use of numerical symbols by the chimpanzee (*Pan troglodytes*): Cardinals, ordinals, and the introduction of zero. *Animal Cognition* **4**, 193–199 (2021).

34 Merritt, D.J. & Brannon, E.M. Nothing to it: Precursors to a zero concept in preschoolers. *Behavioural Processes* **93**, 91–97 (2013).

35 Merritt, D.J., Rugani, R. & Brannon, E.M. Empty sets as part of the numerical continuum: Conceptual precursors to the zero concept in rhesus monkeys. *Journal of Experimental Psychology: General* **138**, 258–269 (2009).

36 Howard, S.R., Avarguès-Weber, A., Garcia, J.E., Greentree, A.D. & Dyer, A.G. Numerical ordering of zero in honey bees. *Science* **360**, 1124 (2018).

37 Cipolotti, L., Butterworth, B. & Warrington, E.K. From 'One thousand nine hundred and forty-five' to *1000,945*. *Neuropsychologia* **32**, 503–509 (1994).

38 Benavides-Varela, S. et al. Zero in the brain: A voxel-based lesion-symptom mapping study in right hemisphere damaged patients. *Cortex* **77**, 38–53 (2015).

39 Devlin, K.J. *Finding Fibonacci* (Princeton University Press, 2017).

40 Davis, H. & Pérusse, R. Numerical competence in animals: Definitional issues, current evidence and a new research agenda. *Behavioral and Brain Sciences* **11**, 561–579 (1988).

41 Hedrich, R. & Neher, E. Venus flytrap: How an excitable, carnivorous plant works. *Trends in Plant Science* **23**, 220–234 (2018).

42 Nieder, A. *A Brain for Numbers: The Biology of the Number Instinct* (MIT Press, 2019).

43 Whiteley, M., Diggle, S.P. & Greenberg, E.P. Progress in and promise of bacterial quorum sensing research. *Nature* **551**, 313–320 (2017).

44 Roitman, J.D., Brannon, E.M. & Platt, M.L. Monotonic coding of numerosity in macaque lateral intraparietal area. *PLOS Biology* **5**, e208 (2007).

45 Testolin, A., Dolfi, S., Rochus, M. & Zorzi, M. Visual sense of number vs sense of magnitude in humans and machines. *Scientific Reports* **10**, 10045, doi:10.1038/s41598-020-66838-5 (2020).

46 Giaquinto, M. Philosophy of number. In: Cohen Kadosh, R. & Dowker, A., eds. *The Oxford Handbook of Numerical Cognition* (Oxford University Press, 2015).

47 Tegmark, M. *Our Mathematical Universe: My Quest for the Ultimate Nature of Reality* (Vintage, 2014).

48 Siniscalchi, M., d'Ingeo, S., Fornelli, S. & Quaranta, A. Are dogs red-green colour blind? *Royal Society Open Science* **4**, 170869, doi:10.1098/rsos.170869 (2017).

49 Tillmann, B. et al. Fine-grained pitch processing of music and speech in congenital amusia. *Journal of the Acoustical Society of America* **130**, 4089–4096 (2011).

50 Butterworth, B. *Dyscalculia: From Science to Education* (Routledge, 2019).

51 Rilling, M. in *The Development of Numerical Competence: Animal and Human Models Comparative Cognition and Neuroscience* (eds S.T. Boysen & E.J. Capaldi) (LEA, 1993).

52 Locke, J. *An Essay Concerning Human Understanding*, ed. J.W. Yolton (J.M. Dent, 1961; originally published 1690).

ACKNOWLEDGEMENTS

My friend Randy Gallistel has been an inspiration for much of what I have written, as should be clear, and he has tried to keep me on the scientific straight and narrow, especially on matters of animal navigation. He has also carefully read and corrected several chapters. Nevertheless, I am sure he would find much to disagree with in the final version, and I look forward to an opportunity to air these disagreements over breakfast, lunch and/or dinner. Bob Reeve has read all the chapters and offered excellent advice which I have tried to follow. Fiona Reynolds has corrected errors in the text of several chapters. Dora Biro and Rosa Rugani contributed their expertise to the chapter on birds. Gary Rose carefully read the chapter on amphibians and reptiles; Lars Chittka has clarified the evidence on invertebrates, notably on his special subject, bees. Rochel Gelman offered sage advice on human development. Christian Agrillo first got me interested in fish, and Caroline Brennan and her team were happy for me to collaborate on their project on zebrafish. Sabine Hyland and Jeffrey Quilter were exceptionally patient with my naive struggles with Inca and Maya counting. Sasha Aikhenvald helped with my puzzlement with Amazonian languages. Lars Chittka, Francesco d'Errico, Angelo Bisazza, Randy Gallistel and Sarah Benson-Amram allowed me to quote their emails describing their descent into animal numerical cognition.

The origin of this book was a meeting at the Royal Society in London in 2017 that I organized with Randy and Giorgio Vallortigara. It seemed to me, and the *New York Times*, so interesting that its themes should be made accessible to the general reader. I am eternally grateful to the Royal Society for their generous funding and administrative support. Many of the participants told me that it was the best meeting they had ever attended, and I like to think that they weren't just being polite.

This book would not have happened without the support and advice of my literary agent Peter Tallack of the Science Factory or, of course, my publisher Richard Milner at Quercus, who must have believed that it would make a good book.

Finally, my thanks as always to my partner Diana Laurillard, the wisest of the wise, and to our daughters Amy and Anna who treat my latest excited discoveries about animal counting with tolerant amusement.

IMAGE CREDITS AND SOURCES

p15 Panther Media GmbH / Alamy Stock Photo; p18 Jeff Edwards; p26 Jeff Edwards after Koehler, O., 'The ability of birds to count', *Bulletin of Animal Behaviour* 9, (1950); p34 Brian Butterworth; p55 Brian Butterworth; p59 Jeff Edwards after Cantlon, J., Fink, R., Safford, K. & Brannon, E.M., 'Heterogeneity impairs numerical matching but not numerical ordering in preschool children', *Developmental Science* 10, (2007); p65 Brian Butterworth; p76; Jeff Edwards; p85 Jeff Edwards after Friberg, J., 'Numbers and measures in the earliest written records', *Scientific American* 250, (1984); p87 Jeff Edwards after Friberg, J., 'Three thousand years of sexagesimal numbers in Mesopotamian mathematical texts', *Archive for History of Exact Sciences* 73 (2019); p89 Public domain; p89 Jeff Edwards after Ascher, M. & Ascher, R., *Code of the Quipu: A Study in Media, Mathematics, and Culture* (University of Michigan Press, 1981); p91 Public domain; p94 Public domain; p103 Jeff Edwards after Henshilwood, C.S., D'Errico, F. & Watts, I., 'Engraved ochres from the Middle Stone Age levels at Blombos Cave, South Africa', *Journal of Human Evolution* 57; p105 Jeff Edwards after Clottes, J., *Les Cavernes de Niaux* (Editions du Seuil, 1995); p110, Joordens, J.C.A. et al., 'Homo erectus at Trinil on Java used shells for tool production and engraving', *Nature*; p136 Jeff Edwards after Brannon, E.M. & Terrace, H.S.,

'Ordering of the numerosities 1 to 9 by monkeys', *Science*; p138 Jeff Edwards after Cantlon, J.F. & Brannon, E.M., 'How much does number matter to a monkey (*Macaca mulatta*)?', *Journal of Experimental Psychology: Animal Behavior Processes*; p147 Jeff Edwards after Cantlon, J.F., Piantadosi, S.T., Ferrigno, S., Hughes, K.D. & Barnard, A.M., 'The origins of counting algorithms', *Psychological Science*; p152: Jeff Edwards after Sawamura, H., Shima, K. & Tanji, J., 'Numerical representation for action in the parietal cortex of the monkey', *Nature*; p163 Jeff Edwards; p169 Jeff Edwards after Kilian, A., Yaman, S., Von Fersen L & Güntürkün, O., 'A bottlenose dolphin discriminates visual stimuli differing in numerosity', *Learning and Behavior*; p172 Jeff Edwards after Thompson, R.F., Mayers, K.S., Robertson, R.T. & Patterson, C.J., 'Number Coding in association cortex of the cat', *Science*; p180 Jeff Edwards after Koehler, O., 'The ability of birds to count', Bulletin of Animal Behaviour 9, (1950); p183 Ditz, H.M. & Nieder, A., 'Neurons selective to the number of visual items in the corvid songbird endbrain', *Proceedings of the National Academy of Sciences*; p205 Jeff Edwards after Balestrieri, A., Gazzola, A., Pellitteri-Rosa, D. & Vallortigara, G., 'Discrimination of group numerousness under predation risk in anuran tadpoles', *Animal Cognition*; p208 Miletto Petrazzini, M.E., Bertolucci, C. & Foà, A., 'Quantity discrimination in trained lizards (*Podarcis sicula*)', *Frontiers in Psychology*; p220 Jeff Edwards after Agrillo, C., Piffer, L., Bisazza, A. & Butterworth, B., 'Evidence for two numerical systems that are similar in humans and guppies', *PLoS ONE*; p222 Jeff Edwards after Dadda, M., Piffer, L., Agrillo, C. & Bisazza, A., 'Spontaneous number representation in mosquitofish', *Cognition*; p223 Jeff Edwards after Bisazza, A. et al., 'Collective enhancement of numerical acuity by meritocratic leadership in fish', *Scientific Reports*; p224 Jeff Edwards after Agrillo, C., Dadda, M., Serena, G. & Bisazza, A., 'Use of number by fish', *PLoS ONE*; p240 Jeff Edwards; p251 Jeff Edwards after Agrillo, C., Dadda, M., Serena, G. & Bisazza, A., 'Use of number by fish', *PLoS ONE*; p253 Jeff Edwards after Gross, H. et al., 'Number-based visual generalisation in the honeybee', *PLoS ONE*; p255 Bortot, M. et al., 'Honeybees use absolute rather than relative numerosity in number discrimination', *Biology Letters*; pp282, 283 Zorzi, M., Stoianov, I. & Umilta, C., in *Handbook of Mathematical Cognition* (ed. J.I.D. Campbell, Psychology Press, 2005); pp300, 311 Brian Butterworth.

INDEX